# Principles of
# Fire Prevention

## 2nd Edition

### David Diamantes

DELMAR
CENGAGE Learning

Australia • Brazil • Japan • Korea • Mexico • Singapore • Spain • United Kingdom • United States

## DELMAR
### CENGAGE Learning

**Principles of Fire Prevention, 2e**
**David Diamantes**

Vice President, Career and Professional
Editorial: Dave Garza

Acquisitions Editor: Janet Maker

Managing Editor: Larry Main

Sr. Product Manager: Jennifer A. Starr

Editorial Assistant: Amy Westel

Vice President, Career and Professional
Marketing: Jennifer Baker

Marketing Director: Deborah S. Yarnell

Sr. Marketing Manager: Erin Coffin

Associate Marketing Manager: Shanna Gibbs

Production Director: Wendy Treoger

Production Manager: Mark Bernard

Sr. Art Director: Benjamin Gleeksman

Manufacturing Buyer: Beverly Breslin

Content Project Management:
PreMediaGlobal

Production House/Compositor:
PreMediaGlobal

For product information and technology assistance, contact us at
**Cengage Learning Customer & Sales Support, 1-800-354-9706**
For permission to use material from this text or product,
submit all requests online at **www.cengage.com/permissions.**
Further permissions questions can be e-mailed to
**permissionrequest@cengage.com**

Library of Congress Control Number: 2010932359

ISBN-13: 978-1-4390-5748-3

ISBN-10: 1-4390-5748-6

**Delmar**
5 Maxwell Drive
Clifton Park, NY 12065-2919
USA

Cengage Learning is a leading provider of customized learning solutions with office locations around the globe, including Singapore, the United Kingdom, Australia, Mexico, Brazil, and Japan. Locate your local office at:
**international.cengage.com/region**

Cengage Learning products are represented in Canada by
Nelson Education, Ltd.

To learn more about Delmar, visit **www.cengage.com/delmar**

Purchase any of our products at your local college store or at our preferred online store **www.cengagebrain.com**

Printed in the United States of America
1 2 3 4 5 6 7 14 13 12 11 10

# Dedication

To my daughters, Nicole and Michelle, who have never failed to make me smile.

# CONTENTS

## Chapter 4    FIRE PREVENTION THROUGH THE CODES PROCESS / 70

## Chapter 5    PLAN REVIEW / 89

## Chapter 6    INSPECTION / 104

## Chapter 7    FIRE PROTECTION SYSTEMS TESTING / 134

## Chapter 8    OTHER FIRE PREVENTION FUNCTIONS / 149

## Chapter 9    FIRE PREVENTION THROUGH INVESTIGATION / 159

## Chapter 10   FIRE PREVENTION THROUGH PUBLIC EDUCATION, AWARENESS, AND THE PUBLIC FORUM / 172

# APPENDICES

# PREFACE

## INTENT OF THIS BOOK

*Principles of Fire Prevention* serves as a comprehensive resource for students at fire departments and fire academies as well as those enrolled in fire science programs at 2- and 4-year colleges. It is intended for a course on fire prevention and adheres to the national *Fire Prevention* course objectives developed by the Fire Emergency Services Higher Education (FESHE) committee at the National Fire Academy. A correlation guide that cross-references the FESHE course objectives with the content in this book can be found on page [xv].

## ORGANIZATION OF THIS BOOK

Understanding why enables students to interpret and apply complex regulations that have evolved over time. Remembering past catastrophic failures enables students to understand the logic behind a particular system or method and to better "sell" the concept to a public we attempt to protect. In response to this need, this book recognizes the importance of fire prevention programs throughout the United States, providing both historical and modern-day examples.

**CHAPTER 1:** The Basis for Fire Prevention, describes the conditions that led to the call for the fire prevention programs we have today.

**CHAPTER 2:** Public Fire Prevention Organizations and Functions, describes the agencies, structures, and functions of government organizations involved in the field of fire prevention.

**CHAPTER 3:** Private Fire Protection and Prevention Organizations, describes the efforts of nongovernmental organizations that may or may not be organized for profit. Within the chapter, I have attempted to dispel the myth that fire prevention was conceived and nurtured by the American fire service; nothing could be farther from the truth. Early attempts to promote the installation of automatic sprinklers systems were opposed by fire chiefs, fearing they would be put out of business. Fire prevention in the nineteenth century was, for the most part, championed by business interests. Their goal was improving the bottom line, but within the boardrooms, there was also an ethic of thriftiness and patriotism mixed with a belief that industry, commerce, and trade were divinely inspired and that the threat of conflagration must be subdued.

**CHAPTER 4:** Fire Prevention Through the Codes Process, discusses model codes and the code process. Fire prevention was embraced by the fire service in the last half of the twentieth century. Fire service influence in the code process has increased in the Twenty-first Century and at no time has there been greater potential for fire service influence. Largely ignored by many in the fire service, no other single issue will have a greater impact on firefighter safety in the future.

**CHAPTERS 5, 6, AND 7:** Discuss the traditional regulatory aspects and functions associated with fire prevention, Plan Review, Inspection, and Fire Protection Systems Testing, respectively. Chapter 8, Other Fire Prevention Functions, covers fire prevention efforts and programs that are not usually considered "traditional," such as construction regulation, enforcement of residential property maintenance codes, and the enforcement of environmental laws and regulations. Chapter 9, Fire Prevention Through Investigation, describes the investigation process from the fire scene to the courtroom, identifies state and federal agencies involved in the process, describes the use of statistics, and discusses the importance of training and adequate equipment.

**CHAPTER 10:** Fire Prevention through Public Education, Awareness, and the Public Forum, makes the case for "selling" fire prevention and underscores the long, rich history of effective public education programs in the United States. Significant statistical evidence suggests that fire prevention programs are most effective when they are embraced and accepted by both the regulators and the regulated. Educating or "selling" fire prevention to the welder, the clerk, the manager, the nurse, the principal, the waiter, the mechanic, the accountant, and the minister means explaining the value and logic of fire prevention. For a program to be effective, the end result must be a public that wants to prevent something that they probably think will not happen where they live or work anyway—fire. Fire happens to somebody else.

**CHAPTERS 11, 12, AND 13:** Fire Prevention Records and Record Keeping, Personnel, and Financial Management, respectively, describe key functions that make or break any program and are often overlooked.

## FEATURES OF THIS BOOK

This book includes many features to help enhance learning for students:

- **NOTES** integrated throughout the chapters highlight critical points for the students to focus on and offer an additional source for review.

- **DISCUSSION QUESTIONS** are included at the end of each chapter, in addition to the Review Questions, to provide topics for classroom discussion.
- **CHAPTER PROJECTS** relate to the chapter and can be used as activities in the classroom or for individual study and allow the students to apply what they have learned.
- **ADDITIONAL RESOURCES** at the end of each chapter list helpful reading materials and Web site addresses in which additional information on the subject can be found.

## NEW TO THIS EDITION

In efforts to keep firefighters appraised of the changes in the world of fire prevention since the publication of the first edition, the author has meticulously researched and incorporated the following revisions to the book:

- Updated fire statistics and analysis
- Recent significant fire incidents and their impact on fire and construction codes and national, state and local fire prevention programs
- Updated technical and code terminology
- The impact of the FESHE and National Professional Development Model (NPDM) on fire and emergency services
- The expanding role of third-party engineering and inspection agencies in municipal code enforcement
- Changes in the model code development process and voting procedures
- Professional certification of fire protection engineering technicians through the National Institute for Certification in Engineering Technologies (NICET)
- New trends in public education programs

## SUPPLEMENT TO THIS BOOK

*Instructor Resources* on CD-ROM is available to accompany the second edition of *Principles of Fire Prevention* and includes the following instructor tools:

- **Lesson Plans and Answers to Questions** in Microsoft Word® highlight the important points in each chapter and correlate to the corresponding PowerPoint Presentations. Answers to Review and Discussion Questions are also provided for easy reference.
- **PowerPoint Presentations** follow the corresponding Lesson Plans and include photos and graphics to enhance classroom presentations.
- **Computerized Testbanks** in ExamView 6.0 are available for each chapter. Testbanks are editable to allow instructors to revise questions or create new tests.
- **FESHE Correlation Grid** includes a cross-reference of the Fire Prevention course outcomes developed by the FESHE committee at the National Fire Academy and the content in the book.

*(Order#: 1-4390-5747-8)*

# FROM THE AUTHOR

Since retiring from the fire department in 1998, I have spent the better part of my time training firefighters and fire inspectors in fire codes and fire prevention. In all honesty, there are few subjects as grisly. Most students initially dread the thought of the subject, but many, if not most, have a change of heart during the training. Student evaluations have consistently identified an aspect of the training that has made it bearable, if not enjoyable—the story behind it all. The who, when, where, and why have typically been disregarded, glossed over, or forgotten by many in our trade. In addition to describing our system of fire prevention, I wanted to make *Principles of Fire Prevention* a good read in the hope that an interested student is a better learner.

Thirty-year-old memories of sitting at a desk in a community college classroom, fighting the urge to doze off, and wondering why I would want to prevent fires, when I really liked going to them, are as vivid as yesterday. To fire science students, I feel your pain. I hope your instructor is successful in selling you on the subject and making you realize that the prevention of fire is your first duty and is in your best interest. The safety of the public you protect, the safety of your brother and sister firefighters, and the economic well-being of our nation are all threatened by fire.

Dave Diamantes
Berryville, Virginia

# ABOUT THE AUTHOR

**David Diamantes** is a consultant, code trainer, curriculum developer, and author. He retired from the Fairfax County Fire and Rescue Department in 1998, as a Captain II, after 25 years of service as a firefighter, company officer, and fire prevention officer. He served as the Supervisor of the Inspections and the Fire Protection Systems Testing Branches and served as the Fire Prevention Division's budget coordinator for several years.

He served as an adjunct instructor at the National Fire Academy, Emmitsburg, Maryland; the Virginia Fire Marshal Academy; Maryland Fire and Rescue Institute; and Harrisburg Area Community College, Harrisburg, Pennsylvania. He is the author of *Fire Prevention: Inspection and Code Enforcement*, 3rd edition, Delmar Cengage Learning, 2007.

He holds an associate's degree in applied science in Fire Science Administration/Fire Investigation from Northern Virginia Community College, Annandale, Virginia, and a bachelor of science degree in Fire Science from the University of Maryland University College, College Park, Maryland. His professional certifications include Certified Building Official (CBO) from the International Code Council; Fire Inspector and Fire Investigator from the Commonwealth of Virginia; and Building Inspector and Plan Reviewer from the Commonwealth of Pennsylvania.

He is a life member of the Virginia Fire Prevention Association and maintains memberships in the International Code Council, International (ICC), Virginia State Fire Chiefs Association (VFCA), and International Association of Firefighters (IAFF).

# ACKNOWLEDGMENTS

George Orwell is claimed to have said, "Writing a book is a horrible, exhausting struggle, like a long bout of some painful illness. One would never undertake such a thing if one were not driven on by some demon whom you can neither resist nor understand." The quote was attributed to him after his death. If he didn't say it, he should have. I have said it several times over the past months, just not as eloquently.

Many people graciously offered information and guidance in this work, and for that I am extremely grateful. Special thanks go to my developmental editor, Jennifer Starr, who survived my questions, prank calls, and complaints. To those I may have inadvertently omitted, I apologize. I'd like to thank the following friends, associates, critics, confidants, mentors, and professionals for their assistance in this endeavor. If there are inaccuracies regarding information they provided, the error is mine.

Jennifer Starr, Janet Maker, Dawn Daugherty, Mark Huth, Jennifer Luck, Alison Pase, Cengage Delmar Learning; Dennis Rowe, New York Board of Fire Underwriters; Mel Cosgrove, ICC; Nancy Cricco, NYU; Dr. Bill Grosshandler, NIST; Tom Herman, Eagle Fire Protection and Consulting; Special Agent Greg Hine, ATF; Mark Johnson, ICC; Maurice Jones, Alexandria Fire Department; Michael Newman, NIST; Duane Perry, Alexandria Fire Department; Donald Mackall, Alexandria Fire Department; Janice Ridenour, U.S. Fire Administration; Jack Rosavear, Sonoma County Fire Department; Steve Sawyer, NFPA; Howard Summers, Chief Fire Marshal, Commonwealth of Virginia (retired); Special Agent Mark Teufert, ATF; Chris Trout, FIREHOUSE Software; Hugh Wood, U.S. Fire Administration; Steve Zenofsky, FM Global; Chief Donald Warner, U.S. Air Force; Scott Boatright, Deborah Showalter, and David J. Thomas, Fairfax County Fire and Rescue Department; and J. Martin Hartman, Fairfax County Fire and Rescue Department, retired.

Both the publisher and I also wish to thank the reviewers who contributed to the development of the book:

Eddie Bain
Illinois Fire Service Institute
Champaign, IL

Dr. Bill Lowe
University of Maryland University College
Adelphi, MD

Ralph DeLaOssa
Long Beach City College
Long Beach, CA

David Murphy
University of North Carolina, Charlotte
Charlotte, NC

Timothy Flannery
John Jay College of Criminal Justice
New York, NY

Larry Perez
New Mexico State University, Dona Ana
Las Cruces, NM

James Goodloe
Florida Division of State Fire Marshal
Tallahassee, FL

Mark Selke
Montgomery College
Rockville, MD

William Hicks
Eastern Kentucky University
Richmond, KY

Eddie O. Smith
Crafton Hills College
Yucaipa, CA

T.J. Johannsen
Bend Fire and Rescue
Bend, OR

W. Faron Taylor
The Overland Group, LLC
Whiteford, MD

Judith Kuleta
Bellevue Community College
Bellevue, WA

Andrew Tinsley
Eastern Kentucky University
Richmond, KY

# FIRE AND EMERGENCY SERVICES HIGHER EDUCATION (FESHE) CORRELATION

In June 2001, The U.S. Fire Administration hosted the third annual Fire and Emergency Services Higher Education Conference at the National Fire Academy campus in Emmitsburg, Maryland. Attendees from state and local fire service training agencies, as well as colleges and universities with fire-related degree programs, attended the conference and participated in work groups. Among the significant outcomes of the working groups was the development of standard titles, outcomes, and descriptions for six core associate-level courses for the model fire science curriculum that had been developed by the group the previous year. The six core courses are Fundamentals of Fire Protection, Fire Protection Systems, Fire Behavior and Combustion, Fire Protection Hydraulics and Water Supply, Building Construction for Fire Protection, and Fire Prevention. The curriculum was updated in February 2008.[1]

## Fire Prevention Course Content

The National Fire Science Curriculum Advisory Committee identified nine desired outcomes involving 12 content areas for the course.[2] This text was written to address each desired outcome within the 12 content areas.

## Course Description

This course provides fundamental information regarding the history and philosophy of fire prevention; organization and operation of a fire prevention bureau; use of fire codes; identification and correction of fire hazards; and the relationships of fire prevention with built-in fire protection systems, fire investigation, and fire and life safety education.

---

[1] 2008 *Fire and Emergency Services Higher Education Conference Final Report*, (Emmitsburg, MD: U.S. Fire Administration) 2008, page 12.

[2] Fire and Emergency Services Higher Education, *Fire Science Model Curriculum* (Emmitsburg, MD: U.S. Fire Administration) 2008, page 10.

# Fire Prevention Core Course—Desired Outcomes

1. Define the national fire problem and main issues relating thereto.
2. Recognize the need, responsibilities, and importance of fire prevention as part of an overall mix of fire protection.
3. Recognize the need, responsibilities, and importance of fire prevention organizations.
4. Review minimum professional qualifications at the state and national level for Fire Inspector, Fire Investigator, and Public Educator.
5. Define the elements of a plan review program.
6. Identify the laws, rules, codes, and other regulations relevant to fire protection of the authority(ies) having jurisdiction.
7. Discuss training programs for fire prevention.
8. Design media programs.
9. Discuss the major programs for public education.

## FESHE Content Area Correlation

The table on the next page provides a comparison of the twelve FESHE content areas within this text.

# FIRE AND EMERGENCY SERVICES HIGHER EDUCATION (FESHE) COURSE CORRELATION GRID

| FESHE CONTENT AREA | TEXT REFERENCES |
|---|---|
| I. History and Development of Fire Prevention | |
|    A. Public, Federal, State and Private Fire Prevention Organizations | Chapters 1 and 2 |
| II. Organization of a Fire Prevention Bureau | |
|    A. Functions | Chapters 5, 6, 7, 8, 9, and 10 |
|    B. Fire Prevention Duties and Responsibilities | Chapters 5, 6, 7, 8, 9, 10, 12, and 13 |
|    C. Fire Prevention Tools of the Trade | Chapters 5, 6, 7, 8, 9, and 10 |
| III. Building Codes and Fire Prevention | |
|    A. Model Building Codes | Chapter 4 |
|    B. Other Codes | Chapter 4 |
| IV. Fire Codes and Fire Prevention | Chapters 4, 6, 7, and 8 |
| V. Structural Elements | Chapters 6, 7, and 8 |
| VII. Identification of Hazards | |
|    A. Common versus Special Hazards | Chapter 6 |
|    B. Hazard Types | Chapters 6 and 8 |
|    C. Nonstructural Hazards | Chapters 6 and 8 |
|    D. Deficiencies in Fire Protection Equipment and Systems | Chapter 7 |
| VIII. Abatement and Mitigation of Hazards | Chapters 6, 7, and 8 |
| IX. Fire Investigation | Chapter 9 |
| X. Public Fire Safety Education | Chapter 10 |
| XI. Plan Review | Chapter 5 |
| XII. Report Preparation and Record Keeping | Chapter 11 |

# INTRODUCTION

A comparison of the current programs of fire prevention bureaus and fire departments in the United States with fire departments of the 1950s reflects the fact that fire departments have assumed additional roles within communities across the country. Many former fire departments are now fire and rescue departments. Some have expanded the role of fire prevention bureaus to include hazardous materials and environmental regulation. Some have even changed their name to the Bureau of Life Safety. There is a perfectly good explanation for the increased role of the fire service. No other organization is as flexible.

No other nonmilitary organization recruits and trains individuals to think on their feet and act as a team. If you dispatch fire apparatus to any problem, the firefighters will step up and solve the problem. Who else do you call when a kid's arm is stuck in a vending machine, when a horse falls through ice, or when a burglar gets stuck in a chimney?

There is a certain danger, however, if you assume that safeguarding life is the sole role of the fire department. Heretics have been burned at the stake; driven out of society; or, at best, shunned for their sins. At the risk of being branded a heretic or offending the sensibilities of some in the fire service, the safety of life and the protection of property are the primary goals of the fire service. Property protection is not a crass motive based on greed. The public health and livelihood are very much tied to the preservation of jobs and a healthy and robust economy. Fire poses perhaps the most significant threat to our material well-being as Americans.

The United States historically has had one of the highest fire loss rates of the industrialized world—both in terms of deaths and dollar loss. This unenviable status has perplexed many experts in the fire world. The United States is health and safety conscious in many areas—automobiles, consumer products, food, and medical drugs, to name a few—and has a vast arsenal of technological resources to combat fire. For such a safety conscious and technologically advanced society to be a leader in fire loss is indeed puzzling.[1]

In this text, I have attempted to identify various systems in use within the public and private sectors for fire prevention. I have also tried to provide insight into the origin and background of fire prevention programs. My underlying motive is to promote the cause of fire prevention and make the point that property protection also saves lives.

There is also the aspect of protecting our brother and sister firefighters. Nine firefighters were killed fighting a fire at the Sofa Superstore in Charleston, South Carolina, on July 18, 2007. The 42,000-square-foot warehouse and showroom had not been inspected since 1998, when the city discontinued routine fire inspections in an attempt at cost savings. Post-fire investigation revealed numerous violations and construction without permits, both of which contributed to the tragedy. In the aftermath, the city hired a fire prevention supervisor and plan reviewer. Chief Mike Chiaramonte, part of a team of outside experts, stated:

> The fire could have been prevented if the property had been constructed and maintained in accordance with state and local codes. If this had been done, the additions would have been built to code or taken down. They would have had either solid two-hour separations or been sprinklered. The flammable liquids would have been stored properly. In all likelihood, the fire would not have spread beyond the loading dock and, if the building had been sprinklered, would have been out when firefighters arrived or have been easily on its way to being extinguished and been a simple one-line fire.

No department should have to live with the tragedy that the Charleston Fire Department endured.

In *Building Construction for the Fire Service*, Frank Brannigan consistently made the point that it is not the fire that is likely to kill you; it is the building. Fire service influence in the code process can reduce the number of future firefighter fatalities and injuries, but it will require a commitment from fire service organizations, labor, and you. A reduction in the number of and in the severity of fires through effective fire prevention programs will surely bolster our nation's economy, lower the number of civilian fire deaths and injuries, and reduce the number of firefighter fatalities and injuries.

In 1992, the U.S. Congress created the nonprofit National Fallen Firefighters Foundation (NFFF). Its original mission was to honor fallen firefighters and to assist their survivors in rebuilding their lives. In 2004, the NFFF hosted the first ever Firefighter Life Safety Summit to examine the increasing numbers of firefighter line-of-duty deaths. During the summit, 16 firefighter life safety initiatives were identified as a part of the NFFF's National program to prevent firefighter deaths. The program titled "Everyone Goes Home" is being implemented throughout the country in cooperation with the United States Fire Administration. The goal is to affect a 25% reduction in firefighter deaths over the next 5 years and a 50 reduction in 10 years. Initiatives 14 and 15 directly address fire prevention programs and efforts, and several others impact or are impacted by fire prevention programs.

### 16 Firefighter Life Safety Initiatives

1. Define and advocate the need for a cultural change within the fire service relating to safety; incorporating leadership, management, supervision, accountability and personal responsibility.
2. Enhance the personal and organizational accountability for health and safety throughout the fire service.

3. Focus greater attention on the integration of risk management with incident management at all levels, including strategic, tactical, and planning responsibilities.

4. All firefighters must be empowered to stop unsafe practices.

5. Develop and implement national standards for training, qualifications, and certification (including regular recertification) that are equally applicable to all firefighters based on the duties they are expected to perform.

6. Develop and implement national medical and physical fitness standards that are equally applicable to all firefighters, based on the duties they are expected to perform.

7. Create a national research agenda and data collection system that relates to the initiatives.

8. Utilize available technology wherever it can produce higher levels of health and safety.

9. Thoroughly investigate all firefighter fatalities, injuries, and near misses.

10. Grant programs should support the implementation of safe practices and/or mandate safe practices as an eligibility requirement.

11. National standards for emergency response policies and procedures should be developed and championed.

12. National protocols for response to violent incidents should be developed and championed.

13. Firefighters and their families must have access to counseling and psychological support.

14. Public education must receive more resources and be championed as a critical fire and life safety program.

15. Advocacy must be strengthened for the enforcement of codes and the installation of home fire sprinklers.

16. Safety must be a primary consideration in the design of apparatus and equipment.

## NOTE

1. *Fire Death Rates, an International Perspective* (Washington, DC: United States Fire Administration, 1997), page 1.

2. *Firefighter Life Safety Summit Initial Report* (Washington, DC: National Fallen Firefighters Foundation, United States Fire Administration, 2004), page 4.

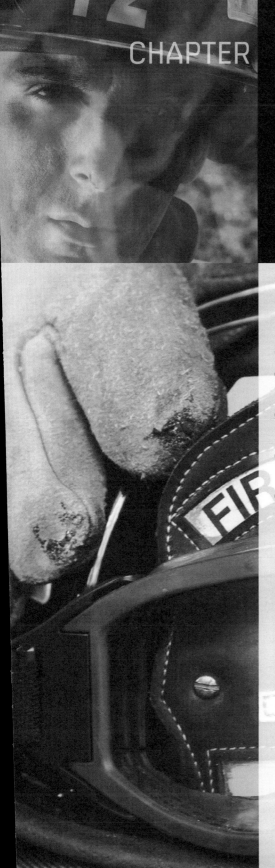

## CHAPTER 1

# The Basis for Fire Prevention

## LEARNING OBJECTIVES

Upon completion of this chapter, you should be able to:

- Discuss the fire problem in the United States and give reasons for its existence.
- Contrast the fire record of the United States with the records of other countries.
- Name organizations that have been instrumental in our nation's fire prevention efforts.
- Discuss the effect that timing has on the adoption and enforcement, or lack of enforcement, of fire prevention regulations.

# THE CALL FOR FIRE PREVENTION—WHEN AND WHY

A look back at efforts to prevent the occurrence of hostile fires and reduce the impact of those that start is largely one of reaction followed by inaction. In the aftermath of a catastrophic fire, elected officials feel the need to act, and a law or several laws are passed. Over time, adherence to the law becomes inconvenient or burdensome, and people tend to forget the incident that led to the law. Fire, after all, usually happens to someone else. Perhaps we are just overly optimistic, but history shows that we are probably more gullible than optimistic when we believe that fire is just a remote possibility.

Fire is not the only thing we believe only happens to someone else. Within a month of the September 11, 2001, terrorist attacks on New York and the Pentagon, there were people interviewed by the news media in airports, complaining about the unnecessary inconvenience posed by increased security measures. Memories fade with the passage of time. The political process is better at dealing with the "now" or the "just happened" than with the future. And the political process has not been particularly effective in dealing with our national fire problem throughout our 200-plus years. Many of our fire prevention successes have been the result of forces outside the political process acting in their own self-interests. In many cases, their successes have benefited the entire country. The insurance industry is the most prominent example of a nongovernmental entity that has had a significant impact on fire prevention through the development of regulations.

## The American Fire Problem

Our national fire record has historically been one of the worst in the Western world. In the 2001 edition of the U.S. Fire Administration's *Fire in the United States*, the problem was summarized in stark language:

> Fire Departments respond to an average of 2 million fire calls each year. This fire problem, on a per capita basis, is one of the worst in the industrial world. Thousands of Americans die each year, tens of thousands of people are injured, and property losses reach billions of dollars. There are huge indirect costs of fire as well—temporary lodging, lost business, medical expenses, psychological damage, pets killed, and others. To put this in context, the annual losses from floods, hurricanes, tornadoes, earthquakes, and other natural disasters combined in the United States average just a fraction of the losses from fires. The public, the media, and local governments, are generally unaware of the magnitude and seriousness of the fire problem to individuals and their families, to communities, and to the nation.[1]

**NOTE** Our national fire record has historically been one of the worst in the Western world.

Unfortunately, the previous edition of *Fire in the United States* contained an almost identical summary, as has almost every report or study of fire protection in U.S. history. In 1971, President Richard Nixon appointed 24 individuals to the National Commission on Fire Prevention and Control. The commission's original report, titled *America Burning,* was released in 1973 and was a significant milestone for fire prevention and protection in the latter half of the twentieth century. The report concluded: "The richest and most technologically advanced nation leads all the major industrialized countries in per capita deaths and property loss from fire."[2]

In May 2000, *America at Risk,* the report of a "recommissioned" America Burning panel, was submitted to the director of the Federal Emergency Management Agency. The commission reported on the nation's progress since the original *America Burning* report in 1973:

> The frequency and severity of fires in America do not result from a lack of knowledge of the causes, means of prevention or methods of suppression. We have a fire "problem" because our nation has failed to adequately apply and fund known loss reduction strategies.[3]

## The Current Trend

Statistics from the U.S. Fire Administration (Figure 1-1) reveal a downward trend in the number of fires and the number of injuries and fatalities from fire. Note that direct dollar loss from fire continues to increase. Indirect loss, particularly business

**FIGURE 1-1**

Statistics reveal a downward trend in the number of fires, and the number of injuries and fatalities from fire (Source: U.S. Fire Administration)

| YEAR | FIRES | DEATHS | INJURIES | DIRECT DOLLAR LOSS IN MILLIONS |
|------|-------|--------|----------|-------------------------------|
| 1999 | 1,823,000 | 3,570 | 21,875 | $10,024 |
| 2000 | 1,708,000 | 4,045 | 22,350 | $11,207 |
| 2001[1] | 1,734,500 | 3,745 | 20,300 | $10,583 |
| 2001[2] | - | 2,451 | 800 | $33,440 |
| 2002 | 1,687,500 | 3,380 | 18,425 | $10,337 |
| 2003 | 1,584,500 | 3,925 | 18,125 | $12,307 |
| 2004 | 1,550,500 | 3,900 | 17,875 | $9,794[3] |
| 2005 | 1,602,000 | 3,675 | 17,925 | $10,672 |
| 2006 | 1,642,500 | 3,245 | 16,400 | $11,307 |
| 2007 | 1,557,500 | 3,430 | 17,675 | $14,639 |
| 2008 | 1,451,500 | 3,320 | 16,705 | $15,478 |

[1] Excludes the events of September 11, 2001.

[2] These estimates reflect the number of deaths, injuries, and dollar loss directly related to the events of September 11, 2001.

[3] The decrease in direct dollar loss in 2004 reflects the Southern California wildfires with an estimated loss of $2,040,000,000 that occurred in 2003. The dollar loss estimate for 2007 includes the California Fire Storm with an estimated property loss of $1,800,000,000. For 2008, the direct dollar loss includes the California Wildfires at an estimated loss of $1,400,000,000.

interruption, which may result in plant and factory shutdowns, layoffs, and sometimes permanent closures or relocations, is often difficult to completely quantify but can be significantly higher.

## THE HISTORY OF FIRE AND FIRE PREVENTION

### Fire in the Early Days

Evidence from anthropological excavations indicates that humans used fire for heat and light sometime around 500,000 BC.[4] Animal bones and charred wood have been found dating to this era, indicating that humans had "captured" fire for their use, probably transferring burning embers from the site of naturally occurring fires to their camps. The ability to capture fire enabled early humans to range farther from the more temperate areas of the earth, opening up new territory for our hunter–gatherer ancestors. New territory meant more game and edible plants and less competition from other humans.

Sometime later, humans developed methods of creating fire through friction, enabling them to start fires at will, again increasing their range and ability to cope with climactic conditions. Anyone who has tried to start a fire with flint and steel or a wooden bow understands the skill and patience involved in primitive fire making. Fire enabled humans to fire pottery; smelt copper and tin; and, finally, make iron. Harnessing the power of fire was a milestone in our evolution from creatures that roamed the savannahs in fear of larger predator animals to modern civilization.

**NOTE** Harnessing the power of fire was a milestone in our evolution.

Even in the earliest of times, there must have been cases in which fire escaped its captors with terrible results. Hostile fires must have erupted within the shelters of early humans, destroying tools, foodstuffs, clothing, and even primitive weapons that provided the only protection from the weather and animal predators. In 500,000 BC, our ancestors could not call the Red Cross for disaster relief—they either starved or froze. The impact of hostile fires on prehistoric civilization as a whole was probably insignificant because tribes or families were small and traveled in bands. There were no cities or towns to burn down. There were no cultural landmarks or large industries to lose.

Things changed when people began to live in close proximity, form cities and societies, and create governments. With the formation of civilization came commerce and trade, and fire became a necessary tool for heat, light, cooking, and industry. Consequently, new methods of creating fire were developed. The lowly match may have been one of the greatest technological inventions in history. Suddenly, fire could be created with the flick of the wrist.

## Technological Progress in Making Fire

Matches are reported to have existed since the time of the Roman Empire, but these were not self-igniting and required heat in order to light.[5] Early friction matches were first made available to the public in the early 1800s and were sold under such prophetic names as "Lucifers" and "Congreves." (Sir William Congreve was the Englishman who invented the military rocket in 1805. It was the "red glare" of Congreve's rockets during the siege of Fort McHenry by the British fleet during the War of 1812 that inspired Francis Scott Key to write the famous words of our national anthem.[6]) As new technologies for fire starting developed, the friction match was replaced by the safety match. However, something else new had developed: when fire escaped from its harness, instead of destroying the campsite and a handful of hapless members of a family, fire could destroy a town or a city. Fire could even be used as a weapon against other cities. Just as today, in the aftermath of fire disasters, society made attempts to prevent their recurrence.

## Fire Prevention in the Early Days

### 300 BC

The first recorded attempts at fire prevention and protection took place in Rome in about 300 BC. Slaves were organized into a combination night watch and firefighting force called the Familia Publica. In about 24 BC, the Roman Emperor Augustus instituted perhaps the first municipal fire department, the Corps of Vigiles, which performed fire patrol and fire-extinguishing duties, with members assigned to specific functions such as water supply or pump operation. Roman law assigned the responsibility for determining cause and origin of a fire to a municipal official akin to a modern fire marshal, permitting corporal punishment for those involved in the ignition of accidental fires and directing the official to deliver "incendiaries," or those involved in the crime of arson, to the Prefect of the City for prosecution.[7]

> **NOTE** The first recorded attempts at fire prevention and protection took place in Rome in about 300 BC.

### AD 1000

Early attempts at preventing fire by regulating public behavior can be traced to England. In 1066, William the Conqueror decreed that all home fires were to be extinguished and covered every evening at a time signaled by the ringing of a bell. The metal cover used to cover the hearth was called a _couvre feu,_ or fire cover. Over time, the term has been changed to "curfew" and has come to mean a time at which persons are to be home and off the streets. In his _Evolution of the Fire Service_, B. J. Thompson remarks that William the Conqueror's _couvre feu_ might not have been that far removed from the intent of today's curfew because without the light from the hearth fires throughout his kingdom, it would have been difficult for his subjects to plot a revolt.[8] England's first curfew can be traced to the city of Oxford in AD 872.

## AD 1500

In the 1500s, English cities passed ordinances regulating bakers and candle makers, two hazardous trades that involved the use of fire in the close quarters that were the early cities. Laws were also enacted that regulated or prohibited wooden chimneys and thatched roofs and that even mandated brick or stone firewalls between buildings. All had differing levels of success, depending on the collective recollection or indifference of the public and governing officials.

## AD 1666

The Great Fire of London occurred in 1666, following on the heels of the bubonic plague or "black death." In an average year at the time, around 17,000 people died in the City of London.[9] However, in 1665, the year preceding the fire, deaths in London attributed to the bubonic plague exceeded 68,000. The fire, which originated in kindling stored near the oven of King Charles's baker,[10] burned for 5 days and nights, destroying 13,200 homes, 87 churches, Saint Paul's Cathedral, 20 warehouses, and 100,000 boats and barges.[11] Nevertheless, it took 2 years for Parliament to enact the London Building Act, and commissioners were not appointed to enforce the regulations for another 108 years in 1774.[12] In the aftermath of the Great Fire of London, a physician named Nicholas Barbon, one of the few who had not fled the city during the plague, formed a group for the insurance of buildings against fire. Shipping insurance had been in existence for some time in England, but property insurance had not. Barbon's effort grew into the London Fire Office, which later led to the formation of a fire brigade to extinguish fires in insured properties, the precursor to the modern fire service.

## Fire in American History

Has modern America developed an indifference to fire in our third century as a nation? Have the success and the bounty that accompanies it caused us to lose the good sense of fire danger that our founders had? The answer is very simple—we suffer from the same indifference that has plagued our nation since its inception. The first permanent English colony in what was to become the United States was established at Jamestown, Virginia, in 1607. Indians who inhabited the area were none too pleased by the arrival of the Englishmen, and they forced the colonists to keep constant vigilance with loaded muskets and water buckets to defend against attacks and fires.

In 1608, fire destroyed most of Jamestown's buildings and provisions, leaving the colonists 3,000 miles from England and surrounded by hostile natives. Captain John Smith later wrote of the fire's impact on the colony: "Many of our old men diseased, and of our new want for lodging, perished."[13] The city of Boston, Massachusetts, was ravaged by nine serious fires before the American Revolution. In response, the Boston general court ordered all buildings to be constructed of brick or stone with slate or tile roofs. The law was never enforced.

Perhaps the most famous conflagration occurred in Chicago in October 1871, when 17,500 buildings were destroyed, more than 300 persons were killed, and more than 100,000 were left homeless.[14] Within 1 month of the fire, Joseph Medill, publisher of the Chicago Tribune, was elected Chicago's twenty-third mayor. Medill ran as the "Fireproof Party" candidate, promising to keep Chicago from ever again suffering the scourge of another conflagration. A powerful Republican journalist, Medill had been an early Free-soiler and a supporter of Abraham Lincoln. His "Fireproof Party" label was proof that even a catastrophe like the great Chicago Fire can be leveraged politically. Within 3 years of the fire, however, reports had reached the National Board of Fire Underwriters (an arm of the insurance industry) that fire safety conditions in Chicago were deplorable.

The board's investigating committee's report on conditions in Chicago revealed that conditions were actually worse than before the fire in 1871. The fire department was "neglected" by the city commissioners and was ill trained and poorly equipped. Fire wardens, who were entrusted with construction inspections, made no effort to enforce the city's building regulations.[15] The insurance industry literally strong-armed Chicago's elected officials into improving the fire department and fire safety conditions. They threatened to cancel every insurance policy within the city. The city was given 3 months to comply—or else. At the insistence of the business community, which clearly understood the potential impact of losing insurance coverage, the city conceded 1 week before the deadline.

In *Safeguarding the Home Against Fire*, a 91-page booklet prepared by the National Board of Fire Underwriters and provided free of charge to more than 2 million schoolchildren in 1918,[16] American "fire waste" was compared with that of European countries[17] (Figure 1-2 and Figure 1-3). We are a nation of proud traditions. The effective prevention of fire apparently has never been one of them.

NOTE  We are a nation of proud traditions. The effective prevention of fire apparently has never been one of them.

## Early Fire Prevention Efforts in the United States

Early fire prevention efforts undertaken by state and local governments, usually in the aftermath of a devastating fire, were in the form of laws or ordinances usually prohibiting the storage or use of flammable materials, limiting hazardous occupations in certain districts, and regulating combustible construction. Peter Stuyvesant, governor of the Dutch colony of Nieuw Amsterdam (later called New York), purchased 250 leather buckets, ladders, and hooks and established a tax on every chimney in the colony in order to maintain the equipment. Stuyvesant also established a roving band of fire

**FIGURE 1-2**
Safeguarding the Home Against Fire was an early example of public education programs by the business community. (Source: Safeguarding the Nation Against Fire, National Board of Fire Underwriters, 1918.)

**FIGURE 1-3**
American "fire waste" for the year 1913 compared with western Europe. (Source: Safeguarding the Nation Against Fire, National Board of Fire Underwriters, 1918.)

| NATION | FIRE LOSS IN 1913 |
|---|---|
| Holland | $ 0.11 |
| Switzerland | $ 0.15 |
| Italy and Austria | $ 0.25 |
| Germany | $ 0.28 |
| England | $ 0.33 |
| France | $ 0.49 |
| United States | $ 2.10 |

wardens armed with wooden rattles to be sounded in the event of fire. The Rattle Watch later became unpopular with the public, who considered them "prowlers more than protectors"[18] (Figure 1-4).

The Massachusetts colony passed a law prohibiting smoking outdoors in 1638, making it the first no smoking law in American history. Similar laws were passed in other colonies.[19] The Pennsylvania legislature passed a no smoking law that applied to the city of Philadelphia. There is no record that shows that the law was ever repealed.[20]

Fire prevention efforts by nongovernmental organizations also occurred in the American Colonies, in many cases with greater success. In 1752, Benjamin Franklin (Figure 1-5) formed the second mutual fire insurance company in America, The Philadelphia Contributionship for the Insurance of Houses from Loss by Fire, known as the Hand in Hand because of its firemark (Figure 1-6), which was attached to the exterior wall of a building to indicate it was insured. The first insurance company, Charleston's

**FIGURE 1-4**
Governor Peter Stuyvesant instituted the "Rattle Watch," fire wardens that carried wooden rattles to sound an alarm of fire. (Courtesy National Museum of Amercian History, Kenneth E. Behring Center, Smithsonian Institution)

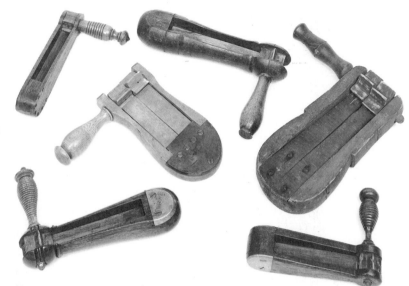

**FIGURE 1-5**
Benjamin Franklin promoted fire prevention and public fire protection. (Courtesy National Museum of Amercian History, Kenneth E. Behring Center, Smithsonian Institution)

**FIGURE 1-6**
The Hand-in-Hand firemark of the Philadelphia Contributionship. (Courtesy of The Philadelphia Contributorship for the Insurance of Houses from Loss by Fire)

Friendly Society for the Mutual Insuring of Houses Against Fire, was founded in 1735 and failed 5 years later after a disastrous fire.[21] Insurance company practices of reducing risk by insuring only well-built and well-maintained properties and periodically inspecting properties in order to upgrade or maintain the level of risk had the effect of improving fire-safe construction and promoting general fire prevention practices.[22] High-risk combustible construction or unsafe practices resulted in prohibitively high premiums or denial of insurance coverage for the property.

As the industrial revolution reached American shores and cities began to grow and become more congested, the threat of **conflagration** became increasingly greater. Investment in industrial buildings and machinery had to be protected by fire insurance. This industry was perhaps the most persistent and energetic in the development of means and methods of preventing fires because its very existence depended on somehow curbing the nation's fire loss. In his 1912 treatise *Fire Prevention and Protection as Applied to Building Construction,* Joseph Freitag, an engineer who specialized in fire protection, mentioned the conflagrations listed in Figure 1-7 and noted:

**FIGURE 1-7**

American conflagrations in the 1800s and early 1900s. (Source: Joseph K. Freitag, Fire Prevention and Protection as Applied to Building Construction, Wiley, New York, 1912, p. 7.)

| DATE | CITY | FIRE LOSS (U.S. $) |
|------|------|--------------------|
| June 1820 | Savannah, Georgia | 3,000,000 |
| December 1835 | New York City | 17,000,000 |
| December 1835 | Charleston, South Carolina | 6,000,000 |
| September 1839 | New York City | 4,000,000 |
| May 1851 | San Francisco | 3,500,000 |
| March 1852 | New Orleans | 5,000,000 |
| July 1866 | Portland, Maine | 10,000,000 |
| October 1871 | Chicago (56 insurance companies bankrupted) | 168,000,000 |
| November 1872 | Boston (65 acres of the city) | 70,000,000 |
| June 1889 | Seattle | 5,000,000 |
| November 1889 | Lynn, Massachusetts | 5,000,000 |
| October 1892 | Milwaukee | 5,000,000 |
| July 1900 | Hoboken, New Jersey | 5,500,000 |
| May 1901 | Jacksonville, Florida | 11,000,000 |
| February 1904 | Baltimore (140 acres, 1,343 buildings) | 40,000,000 |
| April 1906 | San Francisco (earthquake & fire) | 350,000,000 |
| April 1908 | Chelsea, Massachusetts (3,500 buildings) | 12,000,000 |

All large cities contain localities which are pregnant with conflagration possibilities, principally due to the rapid, haphazard growth and construction of such cities. Large areas of wooden buildings may exist, as in San Francisco; or a large store or warehouse, stocked with inflammable goods; inadequately safeguarded, as at Baltimore may provide the cause. The absence of fire walls, shutters or window protection may turn an ordinary fire into one of great magnitude; while such circumstances as low-water pressure, delay in transmitting alarms, bad judgment or disorganization of the fire department, have all been responsible for wide-spread fires.[23]

> **conflagration**   a very large destructive fire that defies control and causes extensive damage over a large area.

## Early Efforts of the Stock Insurance Industry

Early efforts by the fire insurance industry to protect its financial interests involved working toward uniformity in commissions and rates and attempting to standardize regulations within the United States. The industry first attempted to force Congress to develop federal regulations for the industry by reasoning that the sale of fire insurance in the United States was in fact "interstate commerce." In 1866, *Paul v. Virginia* tested the Commonwealth of Virginia's right to impose restrictions on the issuance of a fire insurance policy to a Virginia company from a New York state insurance company. The U.S. Supreme Court ruled that insurance was not interstate commerce; rather, it was the business of the individual states and could be regulated only by the states. This forced the insurance industry to work with each individual state legislature through regional organizations.

It is interesting to note that 75 years later in 1944, the U.S. Supreme Court threw the insurance industry into chaos by reversing *Paul v. Virginia* and overturning state insurance regulations that affected interstate insurance sales. In writing his dissent from the majority, Supreme Court Justice Robert H. Jackson wrote of the decision: "The recklessness of such a course is emphasized when we consider that Congress has not one line of legislation deliberately designed to take over federal responsibility for this important and complicated enterprise."[24]

The insurance industry financed the test case of Samuel Paul in Petersburg, Virginia, through a new organization that had been chartered on July 7, 1866, as the industry and the nation were reeling from a $10 million fire loss in Portland, Maine that occurred 3 days earlier on July 4. It was the first July 4th holiday since hostilities between the states had ended. The fire was caused by a young boy playing with firecrackers in a boat shop, surrounded by flammable materials. Of the $10 million loss, about half was insured.

Rumors circulated that claims would not be paid because the insurance companies had been bankrupted by the large loss. In the wake of the fire, the National Board of Fire Underwriters was formed by resolution of the major **stock fire insurance** companies. The board's mission was to maintain uniform rates and commissions, repress incendiarism and arson, and devise and give effect to measures to provide for the common interests of the group. The board's attempts to maintain uniform rates and commissions proved to be futile and were abandoned over the years; however, the board's accomplishments in other areas were perhaps the greatest of any other group.

**stock fire insurance**  fire insurance provided by commercial, for-profit companies.

## ✳ The National Board of Fire Underwriter's Accomplishments

Books are written about men and women whose actions affect the course of our world, usually through genius, compassion, or bravery and sometimes, unfortunately, through treachery, inhumanity, or evil. But many, if not most, of the events that have had the most significant impact on our world have gone unrecorded and have been long forgotten. The stiff-collared insurance executives shown in Figure 1-8 are in large part responsible for the system of fire prevention and protection that affects every one of us today. Their mark is on the underpinnings of our modern system of construction and fire safety codes, municipal water supply, fire apparatus, municipal fire alarm systems, fire departments, and the fire insurance rating system. Rarely will you hear the activities of the board mentioned during discussions of the evolution of the fire service.

**NOTE**  Rarely will you hear the activities of the board mentioned during discussions of the evolution of the fire service.

Although the board failed at its original mission of maintaining rates and commissions, the board's other activities were substantial. It began funding rewards for the conviction of arsonists and developed guidelines for municipal water supplies and firefighting apparatus that evolved into today's standards. The National Fire Protection

**FIGURE 1-8**
Founding fathers of the National Board of Fire Underwriters: Mark Howard, E.W. Crowell, George Hope, and James McLean. (Source: Harry Chase Brearley, 50 Years of Civilizing Force. New York: Stokes 1918, p. 22.)

Association (NFPA) was formed under the auspices of the board in 1896 to promote uniformity in fire protection standards. Underwriters Laboratories (UL) began as the Underwriters' Electrical Bureau, inspecting and testing electrical displays for the 1893 World's Fair in Chicago on behalf of the National Board of Fire Underwriters.[25]

The board first published the *National Electrical Code*® in 1896, began formal surveys of municipal fire departments in the wake of the 1904 Baltimore conflagration, and published the first *National Building Code* in 1905. The board went on to publish building and fire codes as the American Insurance Association, then was absorbed in another reorganization, and became a part of Insurance Services Office, Incorporated (ISO). We will continue to discuss the legacy of the organization formed in the ashes of the Portland conflagration throughout this text because our system of fire prevention and protection has the fingerprints of the board's engineers on every page and on every part.

## Efforts of the Factory Mutuals

While the stock or "for-profit" insurance companies were finding that the most successful method of protecting their financial interests was through the prevention of fire, another group was pursuing a similar path. A system of "factory mutuals" had evolved from small groups of New England cotton mill owners that had banded together in the early 1800s.

In 1835, Zachariah Allen, a mill owner in Allendale, Rhode Island, approached stock insurance companies and requested a discount. He had installed every fire protection feature and appliance available at the time and reasoned that a reduced risk should be awarded. He was informed that rates were determined by an average that represented the class of hazard. "Mr. Allen, a cotton mill is a cotton mill, we average them all together."[26]

Allen organized other mill owners into forming a **mutual fire insurance** company, limited to textile manufacturing. By limiting membership to the best-run mills and requiring each mill to be inspected by an officer of the company each year, risks were reduced. Members of the manufacturer's mutuals were reporting savings of more than 50 percent of the premiums charged by the stock companies.[27]

> **mutual fire insurance**   not-for-profit system in which all policyholders are members of the company; when premiums exceed losses, surplus funds are distributed among the members.

Allen's protégé, Edward Atkinson, became the president of Factory Mutual in 1877. Atkinson is credited with being the first to apply scientific methods to the study of fire causes. When fires were found to have originated within hollow wall cavities where rats nested and lined their nests with discarded matches, Atkinson issued a rule that combustible void spaces would be prohibited within heavy timber "mill" construction buildings.[28] To this day, combustible voids are prohibited in Type IV Heavy Timber construction. Atkinson is also credited with having developed the tin-clad fire door, which is basically unchanged today. Atkinson was also a staunch abolitionist and is said to have helped finance John Brown's raid on the federal arsenal at Harper's Ferry, Virginia, with considerably less success.

By the end of the century, the Associated Factory Mutual Companies hired an engineer and inspector to relieve the company officers of inspection duties, and the Factory Mutual System we know today as FM Global was born. Factory Mutual engineers continue to perform fire risk reduction and prevention inspections for their clients, the Allendale Insurance and Arkwright and Protection Mutual Insurance companies. They work on behalf of the companies that insure the properties, not the companies that own the properties.

## FIRE PREVENTION TODAY

The prevention of hostile fires, the reduction of deaths and fire-related injuries, and the elimination of property losses to fire are of interest to all of us. We all basically want the same thing. Why we want it depends on our individual interests. How we approach the issue and what methods we are willing to use to reach the goal vary greatly. The methods enlisted by governments, the business community, and other organizations with an interest in fire prevention vary, depending on the political and economic climate.

During World War II, insurance company inspectors detailed to the National Bureau of Industrial Protection submitted more than 63,000 inspection reports to government war agencies and departments, noting safety and security deficiencies in private industrial facilities involved in war production.[29] The fear of sabotage and its impact on war production created an environment in which the American people and U.S. industry were willing to submit to intense government scrutiny. All they stood to lose was everything. With the dawn of the twenty-first century came renewed concern and vigilance that our national infrastructure is susceptible to sabotage through arson or bombing. The same axioms that were used to guide the inspectors of the National Bureau of Industrial Protection are valid today.

**NOTE** The same axioms that were used to guide the inspectors of the National Bureau of Industrial Protection are valid today.

Governments generally want to promote public welfare; however, government efforts are limited by public sentiment because government officials are ultimately accountable to the voters. Government intrusion, then, is limited to what level the public is willing to bear, and Americans have little tolerance for intrusive government.[30] In the aftermath of a significant fire loss, the public may be willing—in fact, may be eager—to allow increased government intrusion into their activities. This willingness fades over time.

The interests of the business community in fire prevention goes directly to the bottom line. Many people have looked down on this motive as coarse and perhaps even inhuman. This short-sighted view completely misses the fact that some of the greatest successes in protecting property and saving lives have evolved from the efforts of businesses to protect their financial interests. *Protecting property saves lives.* Every American's well-being is dependent on a robust economy. Hostile fire is the enemy of a robust economy.

**NOTE** The interests of the business community in fire prevention goes directly to the bottom line.

**NOTE** Some of the greatest successes in protecting property and saving lives have evolved from the efforts of businesses to protect their financial interests.

## SUMMARY

Our national fire record has historically been one of the worst in the Western world. Every major study of the U.S. fire problem since the *America Burning* report was released in 1973 has included the same indictment:

> Annual losses from floods, hurricanes, tornadoes, earthquakes, and other natural disasters combined average just a fraction of the losses from fires. The public, the media, and local governments are generally unaware of the magnitude and seriousness of the fire problem to individuals and their families, to communities, and to the nation.[31]

Yet we continue to preach fire prevention during only 1 week in October, and we have failed to accurately report fires' actual toll on the economy and the lives of the American people. Few people see past the loss of life or the injuries.

The problem is not a new one. Our forefathers had the very same problem, perhaps to a greater degree. The abundance of lumber as a cheap building material combined with the desire to build quickly and with few restrictions resulted in overcrowded, poorly planned cities that were targets for massive conflagrations. As we matured as a nation, building codes were developed and adopted, and cities installed extensive water systems, established fire departments, and enforced building and zoning regulations. In many cases, these improvements were literally forced down the throats of elected officials by the insurance industry.

Although we have made many improvements, we have failed to implant the concept of the prevention of fire as an individual's obligation to the community. We have failed to convince elected officials, the media, and the public that we all pay a price for every fire. Increased insurance premiums for the entire community, loss of jobs, and erosion of the tax base are all costs associated with hostile fires. We all pay for our neighbor's fire. We have a right to insist that every person and every business make the prevention of fires an everyday priority.

## REVIEW QUESTIONS

*TEST*

*PG 2-3* 1. Name three national reports on fire protection and prevention in the United States in the latter twentieth and early twenty-first centuries.

*STOCK* 2. What is the term used to describe for-profit insurance companies? *PG 11*

*MUTUAL* 3. What is the term used to describe insurance companies that are formed by groups *PG 13* as not-for-profit entities?

*PG 12* 4. Which group was responsible for the development of the National Building Code in 1905?

*PG 14* 5. When engineers from FM Global (Factory Mutual) conduct inspections at industrial facilities, whose interests are they hired to protect?

## DISCUSSION QUESTIONS

* 1. *Acceptable risk* is the term used to describe the level of fire risk that the general public is willing to bear at a given time. In the aftermath of a well-publicized fire incident, the level of acceptable risk changes, and the public demands action. List some recent fire events that have sparked a public outcry and discuss what steps were taken in response to public demand.

2. Based on your answer to question 1, which groups (public, business, special interest) were involved in the development of fire prevention strategies, what were their motives, and which were successful?

3. Based on your answer to question 1, were the steps taken to improve public safety meaningful, or were they merely window dressing to mollify the public?

## CHAPTER PROJECT

Using the most recent edition of *Fire in the United States* (available from the U.S. Fire Administration at www.usfa.dhs.gov), prepare a list of talking points to be used by your state legislator to justify additional funding for fire prevention activities in your state. Include the following statistics for your state:

- The fire death rate per 1,000,000 population
- Your state's death rate compared with the national average
- Your state's rank in fire deaths per million population
- Your state's rank in total fire deaths
  Include the following statistics based on national data:
- Age groups within the population most at risk
- Property class that experiences the most incidence of fire
- Property class that experiences the most deaths and injuries
- Property class that experiences the most dollar loss by fire

# ADDITIONAL RESOURCES

In-depth information on many of the subjects discussed in this chapter can be found in the following texts and publications and at these Web sites.

## The U.S Fire Problem and Fire Loss Statistics

*America Burning* (1973), *America Burning Revisited* (1987), and *America at Risk* (2000) are available from the U.S. Fire Administration at www.usfa.dhs.gov.

*Fire in the United States*, U.S. Fire Administration, National Fire Data Center, published each year and available at www.usfa.fema.gov.

*Fire Loss in the United States*, National Fire Protection Association, available at www.nfpa.org. Nonmembers may download various statistical reports in PDF format.

## Fires

Harry Chase Brearley, *Fifty Years of Civilizing Force* (Frederick A. Stokes, 1916).

Robert Cromie, *The Great Chicago Fire* (Rutlidge Hill Press, 1994).

Michael Dineen, *Great Fires of America* (Country Beautiful, 1973).

James Leasor, *The Plague and the Fire*, (McGraw-Hill, 1961).

Paul Lyons, *Fire in America* (National Fire Protection Association, 1976).

## The Fire Insurance Industry

National Board of Fire Underwriters, *Pioneers of Progress* (National Board of Fire Underwriters, 1941).

*Prevention, a Factual Visual History of Property Loss and Control, Including the Role Played by Factory Mutual* (Factory Mutual Engineering Corporation, 1996).

A.L. Todd, *A Spark Ignited in Portland* (McGraw-Hill, 1966).

Dan Yorke, *Able Men of Boston* (Boston Manufacturers Mutual Fire Insurance Company, 1950).

# NOTES

1. *Fire in the United States*, 12th ed (Emmitsburg, MD: United States Fire Administration, National Fire Data Center, August 2001), page 1.
2. *America Burning* (Washington, DC: National Commission on Fire Prevention and Control, May 4, 1973), page 1.
3. *America at Risk* (Emmitsburg, MD: Recommissioned Panel for America Burning, Federal Emergency Management Agency, May 2000), page 15.
4. Ronny J. Coleman et al., *Managing Fire Services* (Washington, DC: International City/County Management Association, 1988), page 4.
5. John Emsley, *The 13th Element* (New York: Wiley, 2000), page 66.
6. Ibid., page 74.
7. Alexander Reid, *Aye Ready!* (Edinburgh, Scotland: George Steward & Company, 1974), page 5.
8. Coleman et al., page 7.
9. James Leasor, *The Plague and the Fire* (London: McGraw-Hill, 1961), page 35.
10. Ibid., page 195.
11. Coleman et al., page 10.
12. Arthur Cote, P.E. and Percy Bugbee, *Principles of Fire Protection* (Quincy, MA: National Fire Protection Association, 1988), page 3.
13. Michael P. Dineen et al., *Great Fires of America* (Waukesha, WI: Country Beautiful Corporation, 1973), page 28.

14.  Ibid., page 12.

15.  Harry Chase Brearley, *Fifty Years of Civilizing Force* (New York: Frederick A. Stokes, 1916), page 42.

16.  *1930 Fire Prevention Yearbook* (Baltimore: Hough and Lawson), page 4.

17.  National Board of Fire Underwriters, *Safeguarding the Home Against Fire* (New York: National Board of Fire Underwriters, 1918), page 9.

18.  Coleman et al., page 10.

19.  Coleman et al., page 8.

20.  Ibid., page 10.

21.  Nicholas B. Wainwright, *A Philadelphia Story* (Philadelphia: William F. Fell, 1952), page 21.

22.  *Prevention, a Factual Visual History of Property Loss and Control, including the Role Played by Factory Mutual* (Norwood, MA: Factory Mutual Engineering Corporation, 1996), page 4.

23.  Joseph K. Freitag, *Fire Prevention and Protection as Applied to Building Construction* (Wiley: New York, 1912) page 7.

24.  A.L. Todd, *A Spark Lighted in Portland* (New York: McGraw-Hill, 1966), page 219.

25.  Brearley, page 195.

26.  Peter J. McKeon, *Fire Prevention* (The Chief Publishing: New York, 1912), page 1.

27.  Dan Yorke, *Able Men of Boston* (Boston: Boston Manufacturers Mutual Fire Insurance Company, 1950), page 30.

28.  McKeon, page 3.

29.  *On Guard, the Story of an Industry in War* (Washington DC: National Bureau for Industrial Protection, 1946), page 9.

30.  *Fire Death Rates, an International Perspective* (Washington, DC: United States Fire Administration, 1997), page 12.

31.  *Fire in the United States*, page 1.

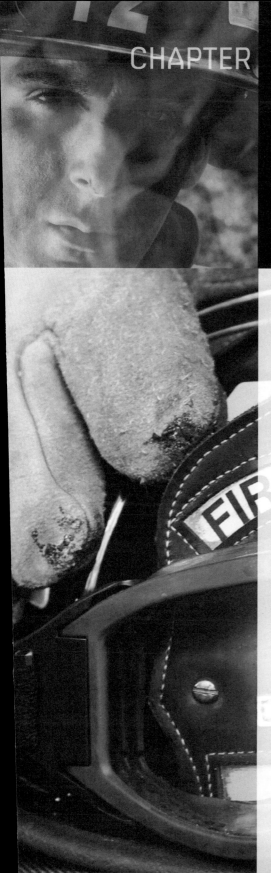

# 2

# Public Fire Prevention Organizations and Functions

## LEARNING OBJECTIVES

Upon completion of this chapter, you should be able to:

- Discuss the role of the federal, state and local governments in the prevention of fires and the reduction in fire deaths and injuries.

- Contrast the roles between the three levels of government in the prevention of fires.

- Name the watershed federal fire programs and describe the events or national conditions that led to their creation.

- List the fire prevention functions performed by traditional fire prevention bureaus and describe nontraditional systems for delivery of those services.

- List federal agencies with fire prevention missions and describe their missions and programs.

# FIRE PREVENTION—MOTIVES BEHIND THE CAUSE

The mantle of fire prevention has been carried by a variety of organizations throughout our history. Some were nobly interested in the public welfare; others were profit driven, looking to reduce insurance payouts. World War II saw a massive fire prevention effort, both to safeguard precious materials needed for the war effort and to protect against sabotage by arson. With access to the natural resources needed to manufacture military vehicles, parachutes, and even life jackets blocked by Germany and Japan, the United States embarked on a massive conservation and recycling effort to marshal the production materials to rearm the military and prepare for war.

"Get in the Scrap" campaigns were aimed at enlisting the public's help in procuring materials such as rubber, silk, cork, and kapok needed for the manufacture of military equipment. Stocks of these precious materials had to be protected from fire. About 18,000 tons of crude rubber, enough for 4 million tires, was lost in a single fire in Fall River, Massachusetts, in 1941.[1] Direct losses from wartime fires in the United States were estimated by the National Fire Protection Association (NFPA) in 1943 at $2 million a day, with indirect losses caused by interruption of production being far more serious for the war effort.[2] Whether for public welfare, as part of the war effort, or as a means to increase profits, we all benefit when the incidence of hostile fire is reduced. Jobs; the economy; and public spending for education, health care, and public infrastructure are all threatened by hostile fires. Those left suddenly unemployed in the wake of a fire lose purchasing power. That loss is felt by the local economy in reduced sales, which eventually leads to further unemployment. Sales tax collection decreases, and unemployed people find themselves unable to meet their financial obligations. Mortgages and consumer loans default, property values drop, and real estate tax collection declines. The reduction in tax collection inevitably leads to reduced public spending for schools and public services.

The fear underlying daily life in the United States in the 1940s, as the nation fought a massive war in Europe and Asia at the same time, must have been strong inducement for every American to do his or her part in preventing fires. The desire for profit has been shown to be one of the most compelling motivators in any society. The not-so-clear motive of doing things in the public interest is not always the most effective. What really is best for the general public is not always clear to elected officials. History is replete with cases in which the political process was not up for the job. Chapter 1's reference to the deplorable fire safety conditions in Chicago 3 years after fire had destroyed 17,000 buildings and a mayor had been elected on the Fireproof Party platform clearly underscores the fact.

Organizations involved in the prevention of fires in the United States fall within two broad categories: those that seek profits and those that do not. Government and not-for-profit organizations that serve the public good seek to improve conditions that affect U.S. citizens. Care must be taken, however, to ensure that carrying out the mission, not perpetuating the organization, is the primary goal. Organizations that are in the fire prevention business and those whose business interests benefit from the prevention of fires play a significant role in the United States. Those who consider the profit motive vulgar and the fire prevention efforts of business for business' sake to be of lesser value to our country need only to look at history for a dose of reality.

Business and industry can claim credit for many of the successes in the field of fire prevention. The roots of our current system of **technical codes** and inspections, public education programs, and fire investigation are in the nineteenth century boardrooms, where men in starched white collars discussed eliminating fire waste and subduing the land in the same meetings.

1890's
AUTOMATED
SPRINKLERS

**NOTE** Business and industry can claim credit for many of our successes in the field of fire prevention.

**technical codes** codes designed to regulate technical processes such as construction; installation of electrical, mechanical, and plumbing systems; hazardous industrial processes; and building, electrical, mechanical, plumbing, and property maintenance codes.

# ORGANIZATIONS WITH THE MISSION OF FIRE PREVENTION

## Traditional Fire Prevention Bureaus

Fire departments in the United States became involved in the prevention of fires in the early twentieth century, although Columbus, Ohio, had inspections by fire companies as early as 1897. The First American National Fire Prevention Convention was held in Philadelphia in 1913. The official record includes the text of every presentation given during the 6-day meeting. Figure 2-1 contains information regarding inspections by municipal fire departments, from a paper entitled *Public Fire Protection*, which was presented on October 14, 1913, the second day of the convention, by Powell Evans and J.S. Mallory. Evans was chairman of the Philadelphia Fire Commission and chairman of the U.S. Chamber of Commerce's Fire Waste Committee. Mallory was acting fire marshal for the City of Philadelphia.[3]

The Fire Department City of New York (FDNY) Bureau of Fire Prevention was established in 1912 as a result of the Sullivan-Hooey Act passed by the New York legislature. The law amended the city's charter and gave control over all the fire department bureaus, including the new Bureau of Fire Prevention, to the commissioner.[4] The Sullivan-Hooey law was passed in the aftermath of the Triangle Shirtwaist fire, in which 146 persons, mostly young women, were killed in a garment factory fire. By 1925, fire prevention bureaus were also operating within the fire departments in the cities of Chicago, Philadelphia, Cincinnati, Detroit, Providence, and Memphis.[5]

**FIGURE 2-1**

Fire department inspection programs as reported during the First National Fire Prevention Conference in 1913. (Source: Powell Evans and J.C. Mallory, "Public Fire Protection Management," Report of the First American National Fire Prevention Convention, 1913, page 151.)

| City | Inspectors | Details |
|---|---|---|
| New York City | 65 inspectors | Tenements were inspected by housing inspectors |
| Chicago | 150 Company officers | Held fire drills in 150 large factories and schools |
| Kansas City | 4 Fire wardens | |
| Rochester, NY | 4 Battalion chiefs | |
| Cincinnati, OH | Company inspections | Began in 1912; half of the commercial buildings were inspected first year |
| Jersey City, NJ | Captains | Authority to issue orders |
| Columbus, OH | Company inspections | Since 1897 |
| Lansing, MI | Captains | Since 1903 |
| Superior, WI | Company inspections | Since 1911 |
| Youngstown, OH | Company inspections | Per capita fire loss reduced from $4 to 40 cents |
| Philadelphia | One man from each of the 80 companies | Each inspector performed five to eight inspections per day |

Early fire prevention bureaus enforced locally developed fire prevention regulations by performing inspections (Figure 2-2). The fire prevention sections of the Greater New York Charter, as amended by the Sullivan-Hooey Act, contained a fairly extensive set of regulations that covered hazards from gaslights and open burning to storage of explosives. The regulations required fire alarms; watchmen in hotels, hospitals, and lodging houses; firefighters in theaters during performances; fire extinguishing equipment; and exit signs. The regulations even made building owners who failed to cover hoist-ways, trapdoors, or fire shutters liable for injuries to firefighters.[6]

The first model fire prevention code was developed by the National Board of Fire Underwriters in 1930. It was similar to current fire codes, regulating hazardous conditions ranging from dry-cleaning operations to vehicle repair garages. The board's *Suggested Fire Prevention Ordinance* was renamed the *National Fire Prevention Code* in 1953. Fire departments and fire prevention bureaus became involved in fire prevention education early in the twentieth century. Fire Prevention Week was inaugurated in 1911 at the suggestion of the Fire Marshals Association of North America. The campaign was led by the National Board of Fire Underwriters and was supported by the U.S. Chamber of Commerce, the NFPA, and others.[7]

**FIGURE 2-2**
Inspection by in-service fire companies has its roots in the late 1800s and early 1900s. (Courtesy of Jeremy Luttrell.)

**NOTE** The first model fire prevention code was developed by the National Board of Fire Underwriters in 1930.

The earliest fire prevention **ordinances** authorized fire prevention bureaus to investigate the causes and origins of hostile fires. Over time, the role and power of many bureaus expanded from cause and origin to follow-up investigation and associated investigative functions, including arrest and referral for prosecution. Many modern fire prevention bureaus have fire department employees assigned to fire investigation duties full time. Many local ordinances charge the fire prevention bureau with the investigation of fires and explosions and crimes related thereto.

**ordinance** law of a political subdivision of a state.

Fire codes require that acceptance tests of required **fire protection systems** be performed in the presence of the fire official. Permits required to install the systems are contingent on submission of plans before installation. Review of fire protection system plans naturally fell to fire prevention bureaus, even though the systems were often required by building codes. With plan review systems already in place, the review of fire access roads and other fire protection features required by model building codes also became the purview of fire prevention bureaus.

**fire protection system** a system that detects fire or combustion products, suppresses or extinguishes fire, retards the passage of fire or smoke, or makes notification or alarm.

## Nontraditional Fire Prevention Bureaus

Not all fire departments have fire prevention bureaus. In some jurisdictions, traditional fire prevention bureau functions simply are not performed or are performed by another agency or agencies. In the aftermath of a fire that killed 25 employees of the Imperial Foods chicken processing plant in Hamlet, North Carolina, the fire chief stated that the entire incident centered around the lack of enforcement of existing codes.[8] The fire department was not adequately staffed to perform inspections.

In some areas, certain traditional fire prevention functions have been shifted to other agencies, usually in the name of streamlining government. Some jurisdictions have located all inspection functions within one agency. The assumption that all types of inspections are basically the same and can be effectively performed by any person with the title "inspector" can have serious consequences if adequate training and supervision are not provided.

> **NOTE** The assumption that all types of inspections are basically the same and can be effectively performed by any person with the title "inspector" can have serious consequences if adequate training and supervision are not provided.

## STATE FIRE PREVENTION AND PROTECTION PROGRAMS

The Commonwealth of Massachusetts empowered a state official to investigate fires and regulate fire waste in 1902. Five other states had followed suit by 1906, and by 1913, 40 states had established the position of fire marshal or other officer with similar powers.[9]

The 50 state governments and territorial governments have fire prevention programs that provide code development, inspection, engineering services, and fire investigation services to the jurisdictions within the state. In many cases, the agency, under the direction of the state fire marshal, provides basic services to rural areas without municipal services. As indicated in Figure 2-3, state fire prevention offices are sometimes

---

**FIGURE 2-3**
The location of the state fire marshall office within state government will affect the duties and responsibilities of the office.

State Police or Public Safety
Department of Insurance
Department of Labor
Department of Housing
Department of Justice
Department of Community Affairs
Department of State
Department of Commerce
State Fire Commission

located within state insurance bureaus, state fire service training agencies, state forestry departments, or state law enforcement agencies. In some states, local fire marshals derive their powers as deputy state fire marshals.

The largest such agency is the California Department of Forestry and Fire Protection (CAL FIRE), which provides fire protection for the state's privately owned wildlands. CAL FIRE also provides fire suppression service under contract for 35 of the state's counties. The California State Fire Marshal's Office is within CAL FIRE; it provides engineering, fire safety education, code enforcement, and investigation services. The fire marshal's office is also responsible for state fire training.

## State Fire Training Programs

State fire training organizations are often not associated with the state fire marshal's office, although they often provide training for the fire marshal's staff, as well as for local fire prevention personnel. Some are located within state universities; others are independent agencies. Training is a fundamental element in any fire prevention program. Without effective training and competent personnel, fire prevention programs cannot reach their full potential.

**NOTE** Without effective training and competent personnel, fire prevention programs cannot reach their full potential.

## FIRE PREVENTION EFFORTS OF THE FEDERAL GOVERNMENT

The federal government has always had an interest in the prevention of fires. For most of our history, federal fire prevention efforts were geared toward the protection of government property and ensuring the continuity of government functions. Protection of the public from fire was, for the most part, an issue for the states and their political subdivisions.

**NOTE** Protection of the public from fire was, for the most part, an issue for the states and their political subdivisions.

Research and fire tests conducted by the federal government, however, had a twofold purpose from early in the twentieth century. Reducing the fire threat to government institutions was a primary consideration in Congress's appropriations for federal agencies that conducted fire research. Obtaining "fundamental engineering data to serve as a basis for the revision and reconstruction of state and municipal building codes" was a stated goal in the establishment of the Bureau of Standards fire research lab in 1914.[10]

In 1906, the total value of buildings under the supervision of the supervising architect of the United States Treasury was $200,000,000. Because the structures were not insured, Congress appropriated funds to establish a materials testing laboratory, aimed at reducing construction costs, while providing the utmost protection against fires and earthquakes. By the end of 1907, the U.S. Geological Survey Structural Materials Testing Laboratory in Saint Louis had conducted 35,000 fire tests on beams, columns, and other structural components.[11] Other government agencies became involved in fire research over time, notably the Commerce Department's Bureau of Standards.

## Catalysts for Federal Fire Prevention Programs

In addition to the protection of the government's infrastructure, other events led the federal government to allocate resources to fire prevention activities. The reported event that led to the establishment of a fire research facility at the National Institute for Standards and Technology was a fire in a pile of leaves on the grounds of the then Bureau of Standards in 1904. In fighting the fire, the employees discovered that fire hoses from different buildings within the complex could not be coupled together because they had different thread patterns.[12] In the same year, the same problem contributed to the Baltimore conflagration in which 140 acres and 1,343 buildings in the downtown area were destroyed.[13] The following year, NFPA, with the active participation of the Bureau, adopted a national standard for hose coupling threads.

President Woodrow Wilson issued the first National Fire Prevention Day Proclamation in 1920. President Warren Harding officially proclaimed the first Fire Prevention Week in 1922 with the statement: "Fire Prevention Week is to be observed by every man, woman, and child, not only during the week designated in this pronouncement but throughout every hour of every day of every year."[14] The idea was not born in Washington, however. Fire Prevention Day had actually been around for almost a decade before its first federal recognition. Fire Prevention Day was first observed in 1911 at the suggestion of the Fire Marshal's Association of North America. The National Board of Fire Underwriters approached state governors, many of whom issued Fire Prevention Day proclamations.[15]

Fire prevention was a big part of the nation's civil defense efforts during World War II. In addition to preserving precious materials for the war effort, the prevention of sabotage by arson was high on the list of national defense concerns. In June 1940, 18 months before the attack on Pearl Harbor, executives of the nation's fire insurance companies organized the Insurance Committee for the Protection of American Industrial Plants to develop strategies to protect American manufacturing facilities that were preparing for war. By June 1941, the committee had established the National Bureau for Industrial Protection in Washington.[16]

**NOTE** Fire prevention was a big part of the nation's civil defense efforts during World War II.

That National Bureau for Industrial Protection first operated with the Federal Bureau of Investigation (FBI) in reducing the threat of sabotage. The FBI used the reports of insurance inspectors to assess security weaknesses at industrial plants. Among the issues evaluated were the fitness of plant guards and watchmen based on their ability and loyalty, as well as fencing, lighting, adequacy of alarms, and facility access. Within months, responsibility for plant security was transferred to the War and Navy Departments, who called on the National Bureau for Industrial Protection to develop regulations for materials and equipment storage. The engineering department of the National Board of Fire Underwriters assigned almost 100 percent of its field staff to the National Bureau for Industrial Protection.[17] Bureau inspectors prepared more than 3,000 individual reports on materials storage for the War Department and War Production Board.[18]

In July 1945, Navy Secretary James Forrestal wrote the National Bureau for Industrial Protection after receiving its final reports:

> I cannot let this occasion pass without expressing my appreciation of the magnificent job which has been done by the National Bureau of Industrial Protection. Sixty-five thousand inspections made by experienced engineers constitute a service to the nation which the Army and Navy would have been at a loss to secure without the aid of your Bureau.

Five months later, President Truman wrote a similarly glowing letter, conveying the country's "grateful thanks" and praise to Harold V. Smith, chairman of the Insurance Committee for the Protection of American Industrial Plants, the parent organization of the Bureau.[19] Whether insurance executives approached the Truman administration in their moment of gratitude is unknown, but no greater token of thanks could have been offered the insurance industry than the Presidential Conference on Fire Prevention.

## President Truman's Fire Prevention Conference

In January 1947, the Truman administration distributed a press release announcing an upcoming national fire prevention conference (Figure 2-4). Representatives of the 48 state governors, business and industry, academia, the fire service and the federal government met in Washington, DC, in May of that year and made remarkable progress. As a result of the conference, 34 governors set up committees on fire prevention. Eighteen states held their own conferences to build on the work begun in Washington.[20]

Among the accomplishments of the conference committees, and perhaps the one with the most significant impact, was the development of a draft model statute permitting the adoption of **model codes**. The National Institute of Municipal Law Officers, Council of State Governments, National Association of Attorneys General, and American Standards Association undertook the presentation of the statute to the state legislatures in 1949. Before development of the model statute and summary adoption by many states, fire and building codes were typically locally developed and sometimes crudely crafted regulations.

**model code**   a code developed by an organization for adoption by governments.

IMMEDIATE RELEASE                                                        JANUARY 3, 1947

## NATIONAL CONFERENCE ON FIRE PREVENTION

For more than a decade the loss of property in the United States due to fires has been steadily mount-
ing year by year. During this period an average of 10,000 persons have been burned to death or have
died of burns annually. In the first nine months of this year fire losses reached the total of nearly half
a billion dollars, with the prospect that final reports for 1946 will show this year to have been the most
disastrous in our history with respect to fire losses.

Additional millions must be added to the nation's bill because of forest fires which, in 1945, accounted
for the destruction of more than 26 million dollars worth of timber, a precious national resource. Also
must be added the enormous sums spent in fighting and controlling fires.

This terrible destruction of lives and property could have been almost entirely averted if proper precautions had been take in time. Destructive fires are due to carelessness or to ignorance of the
proper methods of prevention. These techniques have been tested, but they must be much more
intensively applied in every State and local community in the country.

The President has, therefore, decided to call a National Conference on Fire Prevention, to be held in
Washington within the next few months, to bring the ever-present danger from fire home to all our
people, and to devise additional methods to intensify the work of fire prevention in every town and
city in the Nation.

He has appointed Major General Philip B. Fleming, Administrator of the Federal Works Agency and
of the Office of Temporary Controls, to serve as general chairman of the conference. General Fleming,
who served in a similar capacity during the President's Conference on Highway Safety last May,
already is at work on preliminary arrangements for the meeting, to which will be invited State and
local officials who have legal responsibilities in the matter of fire prevention and control, and representatives of non-official organizations working in this field.

The new impetus given to the prevention of traffic fatalities by the Highway Safety Conference already
has resulted in saving several thousand lives, and the benefits will continue to be felt as the techniques adopted by the conference are increasingly applied. The President is encouraged to hope,
therefore, that a similar attack on fire losses will yield corresponding benefits.

Indeed, that the taking of proper precautions can stem this staggering drain on our resources is well
illustrated in our experience with the Nation's forests. Although the acreage of our unprotected forest
lands amounts to only 25% of the acreage of our protected forests, the losses of the former in 1945
exceeded those of the protected tracts by more than 20%.

The President said: "I can think of no more fitting memorial to those who died needlessly this year in
the LaSalle Hotel fire in Chicago, the appalling disaster at the Winecoff Hotel in Atlanta, and the more
recent New York tenement holocaust than that we should dedicate ourselves anew to ceaseless war
upon the fire menace."

**FIGURE 2-4**

President Truman's 1947 Fire Prevention Conference was attended by forty-eight state representatives,
representatives of the business community, academia, and the federal government.

# FEDERAL AGENCIES INVOLVED IN FIRE PREVENTION

The first major federal program aimed at specifically reducing the fire threat to the general public was instituted in 1974. Public Law 93-478, the Federal Fire Prevention and Control Act, established a federal fire focus while recognizing that fire prevention and protection are fundamentally the responsibility of state and local governments. The result of more than 8 years of hard work and patience, the act was born in a 1966 report, *Wingspread Conference on Fire Service Administration, Education and Research: Statements of National Significance to the United States*. The report called for the establishment of a national commission on fire prevention and control:

> The traditional concept that fire protection is strictly a responsibility of local government must be re-examined. A principle of fire protection which many fire departments and governmental jurisdictions have had to learn the hard way is stated as follows: It is economically unfeasible for any single governmental jurisdiction to equip and man itself with sufficient forces to cope with the maximum situation with which it may be faced.

**NOTE** The first major federal program aimed at specifically reducing the fire threat to the general public was instituted in 1974.

## National Commission on Fire Prevention and Control

The 1968 Fire Research Safety Act established the National Commission on Fire Prevention and Control, a 24-member panel appointed by President Richard Nixon. Its report, *America Burning*, has proven to be one of the most significant forces for fire prevention and protection in U.S. history. Among the findings of the commission were:[21]

- More emphasis on fire prevention is required. Fire departments need to expend more effort on fire safety education, inspection, and code enforcement.
- Better training and education for the fire service is of utmost importance.
- Improved built-in fire protection features in structures would save many lives and avoid property damage.
- Increased involvement of the U.S. Consumer Product Safety Commission (CPSC) in regulation of materials and products affecting fire safety.
- Firefighting, burn prevention and treatment, and protection of the built environment from combustion hazards are important areas of research that have been neglected. Appendix B of this text includes all 90 of the commission's recommendations to Congress.

The commission called for the establishment of the U.S. Fire Administration (USFA), which would establish a national fire data system, monitor fire research, and provide block grants to states and local governments for fire protection and prevention, as well as for the establishment of the National Fire Academy (NFA).[22]

# United States Fire Administration

The USFA was created in 1974 as the National Fire Prevention and Control Administration by the Federal Fire Prevention and Control Act of 1974 (15 USC 2202). In 1979, it was renamed the USFA and became part of the Federal Emergency Management Agency (FEMA). As a result of the Department of Homeland Security Act of 2002, FEMA became part of the U.S. Department of Homeland Security within the Directorate of Emergency Preparedness and Response.

The USFA is headquartered in Emmitsburg, Maryland, and occupies the former campus of Saint Joseph's College (Figure 2-5). The USFA's efforts fall into four basic areas: public education; training for fire and emergency response personnel; fire safety technology, testing, and research; and the collection, analysis, and dissemination of pertinent data.

## USFA Public Education Programs

The USFA develops and delivers educational programs geared toward fire prevention and safety. Public education pamphlets and materials can be obtained through the USFA

**FIGURE 2-5**
The National Emergency Training Center, home of the U.S. Fire Administration and the National Fire Academy.

Publications Center or ordered online. The Fire Safety Directory is a list of materials and resources available from other organizations ranging from burn and scald prevention to electrical hazards. The USFA maintains the list to assist agencies interested in developing public education programs.

**NOTE** The USFA develops and delivers educational programs geared toward fire prevention and safety.

## USFA Training Programs

USFA training programs operate out of the National Emergency Training Center (NETC) located at the Emmitsburg campus (Figure 2-6). The NETC comprises the NFA and the Emergency Management Institute (EMI).

The NFA provides resident training courses at the Emmitsburg campus and courses throughout the country in cooperation with state and regional training organizations. It is estimated that more than 1,400,000 students have received training through a variety of course delivery methods.

Technical and management courses in fire prevention and code enforcement, incident management, hazardous materials, public education, budgeting for fire protection, and emergency planning are among the courses of instruction provided. The number of students who attend courses produced for delivery by other organizations through regional deliveries and NFA-developed handoff courses and through independent self-study is almost five times the number able to attend resident courses.[23]

The EMI focuses on civil defense and natural disaster preparedness. Fires are a natural disaster, and fire service personnel attend the EMI courses in multiagency management of fires, earthquakes, floods, and other natural disasters.

**FIGURE 2-6**
More than 5,000 students attend resident courses at the National Fire Academy each year.

## USFA Fire Safety Technology Programs

The USFA works with public groups and agencies as well as private organizations in promoting fire safety through research, testing, and evaluation. A key issue identified in *America Burning* was firefighter safety and the high rate of firefighter injuries and deaths. The agency develops and distributes research, studies, and other materials to the fire service, design professionals, other fire protection organizations, and the public.

**NOTE** A key issue identified in *America Burning* was firefighter safety and the high rate of firefighter injuries and deaths.

## USFA Data Collection, Analysis, and Dissemination

*America Burning* contained 90 recommendations. The first was for Congress to establish and fund the USFA to provide a national focus for fire protection and prevention issues. The second was that a national fire data system be established to "provide a continuing review and analysis of the entire fire problem."[24]

Lacking valid national statistics, the code development process must rely on anecdotal evidence that may or may not be valid. The National Fire Data Center studies and reports on the nation's fire problem, proposes solutions and priorities, and monitors proposed solutions. *Fire in the United States* is published by the USFA and distributed free of charge.

**NOTE** Lacking valid national statistics, the code development process must rely on anecdotal evidence that may or may not be valid.

## Fire and Emergency Services Higher Education Program

The Fire and Emergency Services Higher Education Program (FESHE) is an NFA program whose mission is to "Establish an organization of post-secondary institutions to promote higher education and to enhance the recognition of the fire and emergency services as profession to reduce loss of life and property from fire and other hazards." To accomplish the mission, annual conferences are held at the NFA campus, where representatives from fire-related degrees programs, state and local fire service training agencies, and national fire service organizations attend. The conferences focus on higher education, sharing ideas, and addressing new challenges (Figure 2-7).

FESHE committees maintain the National Professional Development Model (NPDM), a spreadsheet matrix designed to list professional competencies, education, and training in one document. Training and certification agencies and academic fire programs can adopt the NPDM and customize it to fit their needs, eliminating the often fragmented and stove-piped system of training, higher education, and certification to one that is competency based and completely integrated. FESHE committees also maintain model course outlines for fire-related and emergency medical services (EMS)

**FIGURE 2-7**
The focus of the FESHE conferences is higher education, sharing ideas, and addressing new challenges.
(Source: U.S. Fire Administration)

management degree programs in partnership with publishers to write textbooks used in fire and EMS degree programs.

The National FESHE Bachelor's Model Curriculum includes 15 junior- and senior-level courses developed by the NFA. The NFA has partnered with seven accredited colleges and universities that offer bachelor's degrees with concentrations in fire administration and fire prevention technology. Online training is particularly attractive to fire service personnel whose work schedules make traditional classroom attendance difficult or impossible. It is difficult to gauge the impact of FESHE or the National FESHE Bachelor's Model Curriculum on fire prevention in the United States, but it is safe to assume that without the NFA's training programs, both resident and hand-off, this book would never have been written.

The limited number of positions in fire prevention bureaus leads to the strong probability that many senior chief officers will have little, if any, hands-on experience in fire prevention. The scope and subject matter of the 15 courses included in the model curriculum address the issue head on. The National FESHE Bachelor's Model Curriculum is shown in Table 2-1.

**TABLE 2-1** National FESHE Bachelor's Model Curriculum

| |
|---|
| Analytical Approaches to Public Fire Protection |
| Applications of Fire Research |
| Community Risk Reduction for the Fire and Emergency Services |
| Disaster Planning and Control |
| Fire and Emergency Services Administration |
| Fire Dynamics |
| Fire Investigation and Analysis |
| Fire Prevention, Organization and Management |
| Fire Protection Structures and Systems |
| Fire-Related Human Behavior |
| Managerial Issues in Hazardous Materials |
| Personnel Management for the Fire and Emergency Services |
| Political and Legal Foundations for Fire Protection |
| Issues in Fire/EMS Management |
| Advanced Principles in Fire and Emergency Services Safety and Survival |

# National Institute of Standards and Technology

In addition to identifying the need for a federal fire agency (the USFA) and the establishment of the NFA, the commission made seven recommendations for additional or expanded research by the National Bureau of Standards, a nonregulatory agency within the Commerce Department. The Federal Fire Prevention and Control Act, legislation that resulted from the commission's report, called for the establishment of the Center for Fire Research at the then National Bureau of Standards.[25]

Fire research was not new to the National Bureau of Standards. The Bureau of Standards had been involved in technical fire research since 1914, when Congress funded research on fire-resistant construction materials. Obtaining "fundamental engineering data to serve as a basis for the revision and reconstruction of state and municipal building codes" was a stated goal in the establishment of the Bureau of Standards' fire research laboratory in 1914.[26] The bureau has conducted fire tests and research continuously since that time.

Today, the agency is known as the National Institute of Standards and Technology (NIST), and its Building and Fire Research Laboratory (BFRL) carries on extensive testing and research activities in building materials performance, fire service technologies, fire loss reduction, and other fire-related areas (Figure 2-8).

In the aftermath of the terrorist attacks on the World Trade Center (WTC), investigation of the WTC disaster became a funded program area. Research regarding the effectiveness of building and fire codes, structural fire response, occupant behavior and egress,

**FIGURE 2-8**
Steel from the World
Trade Center under test
at NIST. (Photo courtesy of
the National Institute of
Standards and Technology.)

and aircraft impact damage analysis is being conducted. The BFRL's mission is to "Meet the measurement and standards needs of the Building and Fire Safety Communities."

## National Construction Safety Team Act

On October 1, 2002, President George W. Bush signed the National Construction Safety Team (NCST) Act into law. The law authorized the NIST to establish teams to investigate building failures. The authority is patterned after the National Transportation Safety Board (NTSB) for investigating transportation accidents, and it resulted from the attacks on the Pentagon and WTC complex on September 11, 2001, and the ensuing construction failures.

During congressional hearings on the WTC disaster, witnesses identified critical failures in our system of building design and regulation and the lack of an effective organization to investigate such disasters. Victims' relatives and experts pressed for federal involvement beyond what was legally available in the immediate aftermath of September 11, 2001. The initial federal response was by the Building Performance Assessment Team (BPAT) of FEMA.

During testimony before the U.S. House of Representatives Committee on Science, Professor Glenn Corbett of John Jay College identified failures in the building performance assessment conducted in the wake of the WTC collapse.[27] However, the BPAT lacked the legal authority to conduct an actual investigation. Without legal authority, it was unable to seize and preserve evidence or compel witnesses to provide documents and testify under oath. Representative Sherwood Boehlert, chairman of the House Committee on Science, criticized the conditions that the BPAT members were forced to endure. "We found that the study of the collapse had been hampered by bureaucratic confusion and delay; by a lack of investigative tools; and by excessive controls on the control of information."[28]

With the federal mandate to establish an effective investigative process, the NIST established the NCST, which is patterned after the NTSB. The NCST was pressed into service shortly after The Station nightclub fire in West Warwick, Rhode Island.

**NOTE** The NIST established the NCST, which is patterned after the NTSB.

**NOTE** The NCST was pressed into service shortly after The Station nightclub fire in West Warwick, Rhode Island.

## Bureau of Alcohol, Tobacco Firearms, and Explosives

The Bureau of Alcohol, Tobacco Firearms, and Explosives (ATF) is a law enforcement agency within the Justice Department charged with enforcing federal laws relating to alcohol, tobacco, firearms, explosives, and arson. The agency mission includes regulation, tax collection, and protection of the public. The agency regulates the production and importation of alcohol, ensures that taxes are collected on alcohol and tobacco products, and regulates explosives and firearms importation, manufacture, sales, and storage.

The bureau traces its roots to 1863, when Congress authorized the Treasury Department to hire personnel to serve as revenue agents and reduce the evasion of taxes on distilled spirits. The agency's colorful past includes destroying stills, chasing bootleggers, and successfully prosecuting Al Capone on tax evasion charges. The agency's mission was expanded to include firearms and explosives regulation and enforcement through congressional action. In January 2003, the agency was transferred to the Department of Justice by the Homeland Security Act of 2002 (Public Law No. 107-296).

The agency maintains the Arson and Explosives National Repository, a national collection center for information on arson and explosives-related incidents, and the National Integrated Ballistic Information Network, which provides equipment and support for state and local law enforcement in processing and evaluating gun crime evidence. The ATF's National Laboratory complex opened in 2003. It occupies a 35-acre complex in Ammendale, Maryland. The facility houses the Alcohol and Tobacco Laboratory, the

Forensic Science Laboratory, and the new Fire Research Laboratory. The ATF laboratory system traces its history from 1886, when Congress established a Revenue Laboratory as part of the Department of the Treasury in 1887. Over time, the laboratory's responsibilities expanded to include the forensic analysis of firearms, explosives, fire accelerants, fire devices, and debris from explosives and fire scenes. The Fire Research Laboratory is tasked with conducting fire research and providing case support, training, and education in fire investigation.

The ATF initiates investigations and assists state and local agencies in the investigation of arson and bombings. Through its Certified Fire Investigator (CFI) program, the ATF has worked toward the application of scientific engineering and technology in the field of fire investigation. The ATF maintains a cadre of specially trained agents who are nationally certified to perform fire scene investigation and related law enforcement functions (Figure 2-9). Stationed throughout the country, they are the only investigators trained by a federal law enforcement agency to qualify as expert witnesses in fire cause determination. The ATF uses computer fire modeling as a law enforcement tool, as an aid in the interview and interrogation process, and as a means of refuting the testimony of defense witnesses.

**NOTE** The ATF initiates investigations and assists state and local agencies in the investigation of arson and bombings.

Through its Office of Training and Development, the ATF delivers in-depth training to members of state and local agencies charged with the investigation of arson and

**FIGURE 2-9**
ATF agents work and train with state and local agencies. (Photo courtesy of Duane Perry.)

bombings. Its courses cover investigative and courtroom techniques in the areas of cause and origin, explosives and bombing investigations, and terrorism and explosives. It also offers training for prosecutors on successfully prosecuting arson.

## The National Interagency Fire Center

On October 8, 1871, fire broke out in a barn in Chicago, starting what was to become the most famous conflagration in U.S. history. Three hundred people were killed, and more than 17,000 buildings were destroyed. The same day, a wildland fire burned more than 3,780,000 acres in Wisconsin and Michigan and killed five times as many people, but little is heard about the Peshtigo fire.

The Peshtigo fire was not an isolated incident. A wildland fire 23 years later destroyed Hinkley, Minnesota, and five surrounding towns, killing 418 people in Hinkley alone. There is a mass grave in the town cemetery with a white granite monument commemorating those who were unable to escape on the last train (Figure 2-10). The engineer was forced to run the train through the fire in reverse in order to escape. The town has been rebuilt, including the train station where the fire museum is located. On display are coins melted together that were once loose change in the pocket of a Hinkley resident.

**FIGURE 2-10**
A monument marks the mass grave of the Hinkley fire victims.

Property loss, deaths, and injuries from wildland fires have always been significant problems in the United States. Vast areas of federally owned, undeveloped land spread from coast to coast and border to border. When fires occur and impinge on adjacent communities or threaten federal installations, action must be taken. During the 2002 fire season, approximately 88,458 fires burned almost 7 million acres. Suppression costs for federal agencies amounted to more than $1.6 billion.[29]

In 1965, the National Interagency Fire Center (NIFC) was formed in Boise, Idaho. The center is a cooperative effort among the Bureau of Land Management, Bureau of Indian Affairs, U.S. Forest Service, Fish and Wildlife Service, National Park Service, National Weather Service, Office of Aircraft Services, and National Association of State Foresters. All agencies are members of the National Wildfire Coordinating Group (NWCG), which was created in 1976 by the Secretaries of Interior and Agriculture to facilitate and develop common practices, standards, and training among the organizations.

Statistics from the NWCG indicate that 97 to 98 percent of wildland fires were extinguished during their first burning period, and only 2.5 percent went on to become major disasters.[30] Suppression costs of more than $1 billion annually, and the human toll, including the deaths of 1008 wildland firefighters between 1910 and 2009 and 34 firefighters in 1994, have proven to be more than can be justified. Fire prevention programs involving education, enforcement, and management of fire risk are used to reduce the threat of wildfires.

**NOTE** Ninety-seven to 98 percent of wildland fires were extinguished during their first burning period, and only 2.5 percent went on to become major disasters.

Research shows that wildfires caused by recreational campfires can be reduced by 80 percent through the use of patrols, user contacts, and signage. Fires caused by equipment and children can be reduced by 47 percent.[31] The NWCG agencies conduct public education programs, enforce federal and state fire and open burning laws, conduct fire investigations, and reduce fuel potential through prescribed burns and establishing fire breaks.

**NOTE** Wildfires caused by recreational campfires can be reduced by 80 percent through the use of patrols, user contacts, and signage.

## U.S. Consumer Product Safety Commission

An independent regulatory agency, the U.S. CPSC, was created by Congress in 1972 to protect the public against "unreasonable risks of injuries associated with consumer products." The agency develops standards; conducts research; informs and educates the public; and when necessary, recalls unsafe products. One of the National Commission on Fire Prevention and Control's recommendations in its 1973 *America Burning* report was for the newly created CPSC: "The Commission recommends that flammability standards for fabrics be given a high priority by the CPSC."[32] On November 8, 1974, the CPSC announced a consent order banning the sale of infants' and children's sleepwear that failed to meet the Standard for the Flammability of Children's Sleepwear.

The CPSC has issued recalls on hundreds of unsafe consumer items since that time. In 1998, the first recall of automatic sprinkler heads was announced, and 8.4 million Omega brand fire sprinklers were recalled after resolution of a federal lawsuit in which the CPSC's jurisdiction to issue the recall was challenged. Central Sprinkler Company contended that fire sprinklers were not "consumer items" and were outside the commission's jurisdiction.[33] Since then, the commission has ordered the recall of several problem sprinkler heads.

**NOTE** In 1998, the first recall of automatic sprinkler heads was announced, and 8.4 million Omega brand fire sprinklers were recalled after resolution of a federal lawsuit.

# Department of Housing and Urban Development

The Department of Housing and Urban Development (HUD) has had a significant fire safety impact in residential and health care occupancies through its Minimum Property Standards (MPS). HUD and its predecessor, the Federal Housing Administration (FHA), have maintained the MPS since 1934. The MPS were developed to ensure that properties purchased with federally backed mortgages were constructed to minimum standards for quality, safety, and durability.[34]

# Department of Defense

The fire prevention programs and efforts of the Department of Defense (DOD) are designed to protect the assets of the U.S. military and ensure the ability of its branches to carry out their missions. Unlike a private business concern that could be financially impacted, even to the point of insolvency, the DOD does not get the opportunity to seek reorganization or protection from the bankruptcy court. If a war is lost, so is the country.

**NOTE** The fire prevention programs and efforts of the DOD are designed to protect the assets of the U.S. military and ensure the ability of its branches to carry out their missions.

The U.S. Army, Air Force, Navy, and Marine Corps have fire protection personnel at military installations worldwide. In addition to their fire suppression and crash-rescue duties, they perform fire inspections, develop and deliver fire safety education programs, investigate fires, and serve as fire safety consultants (Figure 2-11). As federal firefighters, they are employed by the organization they protect—a different arrangement from

**FIGURE 2-11**
In addition to their fire suppression and fire prevention duties, Department of Defense firefighters serve as fire protection consultants for the U.S. military. (Photo courtesy of Chief Donald Warner.)

the employment of municipal firefighters. Unlike municipal fire inspectors, DOD fire inspectors are often called on to help fix problems by providing technical assistance.

**NOTE** Unlike municipal fire inspectors, DOD fire inspectors are often called on to help fix problems by providing technical assistance.

With the passage of the National Technology Transfer Advancement Act of 1996, Congress mandated the use of consensus technical standards by federal agencies. In response, the DOD developed Military Handbook 1008, *Fire Protection for Facilities Engineering, Design and Construction*. The document established criteria for DOD installations worldwide, whether on government-owned or leased property. The document incorporated NFPA's *National Fire Codes*, portions of the ICBO *Uniform Building Code*, and Factory Mutual's *Loss Prevention Data Sheets* and other standards.

In April 2003, the *Military Handbook 1008C* was superseded by the *Unified Facilities Criteria, Design: Fire Protection Engineering for Facilities* (UFC). In implementing the UFC, the DOD updated its system of codes and standards to the most current developed by the model code and standards organizations. The UFC is distributed electronically and updated regularly. Updates are effective upon issuance. The Louis F. Garland Fire Academy is the DOD's fire training facility. Located at Goodfellow Air Force Base, Texas, the academy delivers nationally accredited Fire Inspector II and III training for all four military services and the Defense Logistics Agency. The Fire Inspector I course is delivered through correspondence courses. The fire inspector course was originally developed by the Air Force in 1967, and it reduced fire losses Air Force-wide by an estimated 80 percent over the next 10 years.[35] In 1993, DOD adopted the DOD Fire Fighter Certification System, making fire inspector training and certification mandatory for all DOD firefighters.

## Occupational Safety and Health Administration

President Richard Nixon signed the Occupational Safety and Health Act in 1971, creating the Occupational Safety and Health Administration (OSHA). OSHA's mission is workplace safety. In 1970, there were more than 14,000 deaths from job-related injuries, and more than 2.5 million workers were disabled by workplace accidents or conditions. Since 1970, the rate of work-related fatalities has been reduced by half. Brown lung disease has virtually been eliminated in the textile industry.[36] Among other guidelines, OSHA issues standards for fire and explosion hazards and fire brigade staffing, training, and operation. OSHA's respiratory protection regulation 29 CFR 1910.134 (g)(4) is the basis for the two-in/two-out structural firefighting mandate.

## Other Federal Agencies

In addition to those previously mentioned, many federal agencies are either involved in fire research or have extensive fire prevention programs. The National Aeronautics and Space Administration (NASA) has conducted extensive research into fire safety within aircraft, aviation fuels, and other aviation and space-related issues. Nomex, used by the fire service for protective clothing, was the result of NASA research. The Department of State has fire prevention and protection personnel at its embassies worldwide. The Veterans Administration (VA) provides fire suppression and fire prevention services at VA Hospitals across the United States.

## SUMMARY

Fire prevention has traditionally been considered the responsibility of the states and their political subdivisions. Federal involvement has generally been limited to the protection of federal government assets and research. Notable federal fire prevention campaigns have occurred during wartime to protect the defense industry and the national infrastructure from sabotage and disaster wrought by humans.

Watershed events in the area of federal involvement in fire prevention were World War II; President Truman's 1947 Fire Prevention Conference; and the 1973 report of the National Commission on Fire Prevention and Control, *America Burning*. *America Burning* was the catalyst for the creation of the USFA, the NFA, and the National Fire Data Center and for federal focus on fire research for firefighter safety.

## REVIEW QUESTIONS

PG 29    1. What was the name of the 1973 report of the National Commission on Fire Prevention and Control?

PG 34    2. What federal agency operated the Building and Fire Research Laboratory?

PG 35    3. Which federal agency has employees that are trained and certified as certified fire investigators?

PG 21,22,23    4. List four functions performed by traditional fire prevention bureaus.

PG 29    5. What is the name of the federal agency created as a result of the 1973 *America Burning* report?

PG 39    6. Which federal agency is charged with protecting the public from unsafe consumer items?

## DISCUSSION QUESTIONS

1. Of the 90 recommendations of the National Commission on Fire Prevention and Control included in the *America Burning* report (see Appendix B), which three have had the greatest effect on you and your community?
2. In this chapter, specific events have been identified with the passage of laws or the development of fire prevention programs. What recent events could or should have led to new laws or programs? What new laws or programs would you recommend?
3. Is the federal government's role in fire prevention adequate or should more resources be allocated? What would be the effect if Congress were to abolish all federal fire programs and distribute the funds to the states for their fire prevention programs?

## CHAPTER PROJECT

Research the laws of your state regarding the office of the state fire marshal and determine the following:

- The code section that established the position
- Agency and branch of state government to which fire marshals belong
- Duties and powers of the office
- Legal relationship with local fire marshals

## ADDITIONAL RESOURCES

In-depth information on many of the subjects discussed in this chapter can be found in the following texts and publications and at these Web sites.

*America Burning* (1973), *America Burning Revisited* (1987) and *America at Risk* (2000) are available from the U.S. Fire Administration at www.usfa.dhs.gov.

Bureau of Alcohol, Tobacco, and Firearms, Fire Research Laboratory at www.atf.gov/labs.

DOD Firefighters (private Web site operated by Chief Donald Warner) at www.dodfire.com.

National Association of State Fire Marshals at www.firemarshals.org.

National Institute of Standards and Technology, Building Fire Research Laboratory at www.bfrl.nist.gov.

National Interagency Fire Center (wildland firefighting) at www.nifc.gov/index.html.

U.S. Consumer Product Safety Commission at www.cpsc.gov.

U.S. Fire Administration at www.usfa.fema.gov.

The 1966, 1976, 1986, 1996 and 2006 Wingspread Conference Reports are available at www.nationalfireheritagecenter.com/library.

## NOTES

1. *On Guard, the Unsung Story of an Industry in War* (Washington, DC: National Bureau for Industrial Protection, 1946), page 14.

2. *Wartime Fires* (Boston: National Fire Protection Association, 1943), page 1.

3. Powell Evans and J.C. Mallory, *Report of the First American National Fire Prevention Convention*, (Philadelphis: Merchant and Evans, 1914) page 151.

4. Peter Joseph McKeon, *Fire Prevention* (New York: The Chief Publishing Company, 1912), page 26.

5. Percy Bugbee, *Men Against Fire* (Boston: National Fire Protection Association, 1971), page 59.

6. McKeon, *Fire Prevention*, page 30.

7. *Fire Prevention Education* (National Board of Fire Underwriters: New York, 1942), page 11.

8. "Twenty-five Fatality Fire at Chicken Processing Plant, Hamlet, North Carolina, September 1991," Technical Report #057 (Emmitsburg, MD: U.S. Fire Administration, June 1999), page 9.

9. Evans and Mallory, page 16.

10. Daniel Gross, "Fire Research at NBS: the First 75 Years," *Proceedings of the 3rd International Symposium*, Edinburgh, Scotland, G. Cox and B. Langford, eds (Elsevier Applied Science, New York, 1991), page 120.

11. Joseph Kendall Freitag, *Fire Prevention and Fire Protection as Applied to Building Construction* (New York: Wiley, 1912), page 118.

12. Gross, page 120.

13. Freitag, page 7.

14. Center for Safety Education, New York University and the Committee for Fire Prevention Education Representing Eleven National Educational and Fire-Safety *Agencies and Organizations, Fire Prevention Education* (New York: National Board of Fire Underwriters, 1942), page 11.

15. Ibid., page 12.

16. *On Guard, the Unsung Story of an Industry in War*, page 6.

17. A.L. Todd, *A Spark Ignited in Portland, The Record of the National Board of Fire Underwriters* (New York: McGraw-Hill, 1966), page 216.

18. *On Guard, the Unsung Story of an Industry in War*, page 15.

19. Ibid., page 23.

20. *President's Conference on Fire Prevention Final Report* (Washington, DC: Federal Works Agency, 1947), page 4.

21. *America Burning, the Report of the National Commission on Fire Prevention and Control* (Washington, DC: National Commission on Fire Prevention and Control, 1973), page XI.

22. Ibid.

23. U.S. Fire Administration, *FY 2000 Accomplishments* (Emmitsburg, MD: U.S. Fire Administration, 2001).

24. *America Burning*, page 167.

25. Gross, page 127.

26. Gross, page 120.

27. Statement of Professor Glenn P. Corbett, John Jay College of Criminal Justice Before the Committee on Science, House of Representatives, United States Congress, *Learning from 9/11: Understanding the Collapse of the World Trade Center*, March 6, 2002.

28. Stephen Barlas, "NIST and the WTC, Congress Pushes for a Tougher, More Thorough Investigation of WTC Collapse," *NFPA Fire Journal*, September/October 2002.

29. *Wildland Fire Statistics*, National Interagency Fire Center, 2003, page 2.

30. *Wildland Prevention Strategies*, National Wildfire Coordinating Group, 1998, page 9.

31. Ibid., page 10.

32. *America Burning*, page 168.

33. United States Consumer Product Safety Commission, *CPSC Docket 98-2 In the Matter of Central Sprinkler Corp, and Central Sprinkler Co.*, 1998, page 2.

34. National Institute of Building Sciences, *A Study of the HUD MPS for One- and Two- Family Dwellings and Technical Suitability of Products Programs*, March 2003, page i.

35. *Air Force and DOD Fire Academy History*, Donald W. Warner, www.dodfire.com, 2003.

36. Occupation Safety and Health Administration, *All About OSHA* (OSHA 256, 2000), page 7.

# 3

# Private Fire Protection and Prevention Organizations

## LEARNING OBJECTIVES

Upon completion of this chapter, you should be able to:

- Discuss the role of private industry in local, state, and national fire prevention efforts.
- List five industries and professions involved in fire prevention.
- Describe the role of the insurance industry in fire prevention and risk management.
- Describe the role of the design professional in fire prevention and protection.
- Describe the role played by industry trade associations in fire prevention and protection.

*We can do much to shape legislation that will benefit not only our own interests but the whole country, by securing such wise and salutary laws as might prevent the recurrence of other destructive conflagrations.*

Henry A. Oakley (Figure 3-1),
President, National Board of Fire Underwriters, 1873[1]

## PRIVATE FIRE PREVENTION ORGANIZATIONS— PURELY FOR PROFIT?

It is easy to become jaded and question the motives of organizations, particularly organizations that are trying to make money. There are times when motive really does not matter. The good deed, done with perhaps less-than-good intentions, still benefits the recipient. Often, the best interests of business and the public are exactly the same. In the

**FIGURE 3-1**
Henry Oakley called for the insurance industry to promote fire prevention legislation in 1873. (Source: Harry Chase Brearley, Fifty Years of Civilizing Force, Frederick A. Stokes, New York, 1916, p. 52.)

field of fire prevention, public fire prevention programs literally could not exist as we know them without the products and efforts of private sector fire prevention organizations.

**NOTE** Public fire prevention programs literally could not exist as we know them without the products and efforts of private sector fire prevention organizations.

In *Fifty Years of Civilizing Force*, Harry Chase Brearley described the first 50 years of the National Board of Fire Underwriters. Originally organized in 1866 to ensure uniform rates and commissions, the board failed in its original mission but went on to become perhaps the most powerful force for fire prevention and protection the United States has ever known. In describing the board's actions to literally strong-arm the city of Chicago into instituting fire prevention and protection reforms in 1874 under the threat of canceling every insurance policy in the city, Brearley stated: "This was public service of a high order—but, and herein lies its greatest value—its motives were those of practical business, not of altruism."[2]

Private fire prevention and fire protection programs generally fall into three categories: those undertaken by business as part of a risk management system; those that provide fire prevention and protection as a profit-making business service; and those that are not-for-profit, operating in the public interest. Often the roles, products, and missions of the organizations come together and are complementary.

## FIRE PREVENTION RISK MANAGEMENT

Many early fire prevention efforts were undertaken by businesspeople who understood fire risk and intended to reduce the chances of experiencing a catastrophic incident. In 1874, the first practical automatic sprinkler head was patented by Henry S. Parmalee of New Haven, Connecticut.[3] Parmalee was a piano manufacturer, and he developed the sprinkler head for use in his factory. Edward Atkinson, sometimes called the "father of fire protection engineering," was a New England cotton mill owner. Atkinson is credited with developing the tin-clad fire door and advocating the installation of sprinklers in New England mills.

Atkinson was one of the first to view fire prevention as a science. He studied fire causes and fire protection, and his mill became a model for fire prevention and protection. Atkinson went on to become president of the Factory Mutual insurance system, but not without controversy. Most mill owners were skeptical about sprinkler systems and opposed Atkinson's efforts to require their installation as a prerequisite for insurance. One is said to have advised him to "take a sprinkler head with him to the afterlife, for his own protection."[4]

## Corporate Programs

Most large corporations have fire safety and fire prevention programs that are part of the corporation's risk management program. Security is most often under the same umbrella, and more often than not, the security director is also the safety director. In *Fire Safety and*

*Loss Prevention*, a text developed for corporate safety and security officials, Kevin Cassidy identifies the conflicts that often arise between fire safety and security:

> As security/ fire safety director . . . conforming to local laws can and will tie your hands. Security procedures will often be compromised in order to comply with mandated codes. It is crucial that you remind your organization that fire and building regulations are mandated, and compliance is required by law, whereas, most security regulations are not mandated.[5]

Fire prevention bureau inspectors and supervisors are frequently approached by organizations with legitimate security concerns. Fire safety and security can and must be complementary. A system that provides both generally is not the cheapest alternative, and fire department representatives frequently hear complaints that fire or building code requirements simply cannot be met and still provide adequate security. That simply is not true. The truth is that achieving both may be more expensive than simply providing adequate security measures and disregarding fire safety. Building owners, design professionals, and developers want to do it the cheapest way. Fire department personnel need to remember Cassidy's charge to his readers—fire safety regulations are mandated and have the force of law. Security does not. Inside the courtroom, the fact that customers were shoplifting will not justify management's decision to chain the rear exit doors closed. #2

> **NOTE** Fire safety and security can and must be complementary.

> **NOTE** Inside the courtroom, the fact that customers were shoplifting will not justify management's decision to chain the rear exit doors closed.

Fire prevention, fire drills, and **fire brigade** training are all required as part of the overall fire safety plan mandated by model fire codes. The effectiveness of a well-planned and implemented program is starkly outlined in the U.S. Fire Administration's report *Chicken Processing Plant Fires, Hamlet North Carolina and North Little Rock Arkansas*.[6] Both fires occurred in 1991 and involved industrial cooking operations with similar equipment. In the North Carolina fire at Imperial Foods, 25 employees were killed and 54 were injured. The plant owner was imprisoned for code violations, and the plant never reopened after the fire.

> **fire brigade**   a private firefighting force within an industrial or government complex, usually employees of the facility, trained and equipped for firefighting on site.

In the North Little Rock fire at Tyson Foods, all 115 employees were evacuated within 3 minutes, assembled at a predetermined area outside the building, and accounted for by name. There were no injuries, and although Tyson employees were not able to extinguish the fire, plant fire brigade members in self-contained breathing apparatus met firefighters and led them to the fire. Tyson Foods used the downtime to remodel the plant and reopened in 13 weeks. What was the difference between the two incidents? Tyson Foods has a corporate fire safety policy that includes plant inspections, fire drills, and fire brigade training. Inspection reports are forwarded to corporate headquarters monthly. Semiannual fire drills are unannounced in order to simulate realistic conditions. Food products that are on the assembly line must be discarded to meet U.S. Department of Agriculture regulations. Tyson considers the cost as a normal business expense.

# INSURANCE INDUSTRY FIRE PREVENTION PROGRAMS

The insurance industry is responsible for much of our regulatory system used to prevent fires and reduce fire loss. Chapter 1, "The Basis for Fire Prevention," includes a discussion of the early efforts of the stock insurance companies and the mutual companies that evolved among the cotton and woolen mills in New England. The descendents of those organizations exist today, performing many of the same functions as they did in the 1800s.

## Insurance Services Office, Incorporated

The National Board of Fire Underwriters was organized by stock insurance companies in 1866 in the wake of a conflagration in Portland, Maine. The board's original mandate was to regulate rates and commissions in an effort to reduce fierce competition within the industry. Competition led to rate cutting by unscrupulous or incompetent companies. Rock-bottom rates resulted in inadequate cash reserves. Without cash reserves to pay claims, many insurance companies declared bankruptcy in the wake of the conflagrations that plagued the country in the 1800s. Fire victims who were policyholders with the bankrupt companies went unpaid.

The board failed in its original mission of fixing rates but went on to become a huge actuarial or rating bureau and developed effective strategies for reducing fire loss, thereby increasing profits. The board was instrumental in the formation of the National Fire Protection Association (NFPA); established Underwriters Laboratories (UL); and developed and maintained the first model electrical, building, and fire codes in the country. The board developed the *Standard Grading Schedule for Public Fire Departments* in 1916, known today as the **Public Protection Classification System** (PPC). The board became part of the American Insurance Association in 1965 and the Insurance Services Office, Incorporated (ISO) in 1971.

**Public Protection Classification System**   the ISO's system of rating a jurisdiction's system of public fire protection, including the fire department, water supply, and communications system.

The ISO provides statistical analysis and actuarial service, automated information systems that supply information regarding properties across the country, and consulting services. The ISO maintains the PPC (Figure 3-2), formerly the Fire Department Grading Schedule, and in the wake of Hurricane Andrew (1992) and the Northridge Earthquake (1994), the **Building Code Effectiveness Grading Schedule** (BCEGS). Generally, fire

**FIGURE 3-2**
ISO's PPC began with the National Board of Fire Underwriters in the 1800s. (Courtesy of ISO.)

insurance rates are substantially lower in communities with good PPCs.[7] A community's PPC is based on:[8]

#3 ● Fire alarm and communications system
- ○ Telephone system
- ○ Telephone lines
- ○ Dispatching system
- ○ Call center staffing

#3 ● Fire department
- ○ Equipment
- ○ Staffing
- ○ Training
- ○ Station locations

#3 ● Water supply
- ○ Condition and maintenance of hydrants
- ○ Available fire flow

**NOTE** The ISO maintains the PPC, formerly the Fire Department Grading Schedule.

**Building Code Effectiveness Grading Schedule**  a system developed by ISO to evaluate the potential effectiveness of local government building regulatory systems, patterned after the public protection classification system (fire department grading schedule).

For many years, fire chiefs have used the PPC program as a benchmark for justifying staffing, equipment, and improvements to the municipal water supply. Lower insurance rates make a community attractive to businesses. Increased property tax collection from businesses reduces the tax burden on homeowners.

The years 1992 and 1994 were two of the worst on record for catastrophic losses from natural disasters. In the aftermath of Hurricane Andrew and the Northridge Earthquake, many demanded stricter building regulations. In-depth surveys of the damage indicated that inadequate construction codes were not the root cause of the widespread damage. Studies revealed that lax enforcement of building codes was responsible for a considerable amount of the damage to structures. A study by Factory Mutual Insurance Group reported that damage to structures from Hurricane Andrew could have been reduced up to 55 percent if codes had been strictly enforced.[9] California's Governor Pete Wilson noted in his report to the California Seismic Safety Commission:

> Although statistically rigorous data is not available to establish even rough percentages of damage resulting from a lack of code conformance, there is ample evidence that failure to follow code requirements in design, lax plan review, and shoddy construction inspection resulted in significant damage to conventional homes.[10]

**NOTE** Damage to structures from Hurricane Andrew could have been reduced up to 55 percent if codes had been strictly enforced.

In response, a system for measuring the effectiveness of construction regulation was developed by the ISO. The program was instituted in 1995 and was gradually implemented in all areas of the country by 1999. The BCEGS is based on:

- Code administration
  - Use of a model building code and edition
  - Local modifications
  - Zoning provisions
  - Training and certification of employees
  - Contractor and builder licensing
- Plan review
  - Staffing levels
  - Staff qualifications
  - Level of detail of review
  - Staff performance evaluations
- Field inspection
  - Staffing levels
  - Staff qualifications
  - Level of detail of inspections

- ○ Staff performance evaluations
- ○ Final inspections
- ○ Issuance of Certificate of Occupancy

The development of the BCEGS has resulted in more effective construction regulation, which will certainly result in reductions in fire loss over time. The primary purpose of a fire code is to prevent hostile fires and explosions, protect people through proper planning and preparation, and reduce fire spread potential through the regulation of fuels and the maintenance of fire protection features. The primary purpose of a building code is to prevent collapse by managing the impact of loads— loads that act on structures include gravity, wind, and snow; live loads such as people and furnishings; dead loads such as built-in equipment; and loads such as fire or earthquakes. Fire codes are aimed at fire prevention. Building codes aim to manage the impact of fire.

**NOTE** Fire codes are aimed at fire prevention. Building codes aim to manage the impact of fire.

ISO products for insurance underwriting include databases with information on properties and hazards that assist insurance underwriters. FireLine is a computer tool for evaluating the risk of exposure from wildfires. The program combines known risk factors for fuel, slope, and fire department access with satellite imagery.

## New York Board of Fire Underwriters

The New York Board of Fire Underwriters was one of the many local boards that represented stock insurance companies and cooperated with the National Board of Fire Underwriters. In 1881, Thomas Edison wrote a letter to the New York Board of Fire Underwriters (Figure 3-3). An article had appeared in the *New York Evening Post* referring to a resolution of the board's requesting information on a new fire hazard, electricity.

Edison stated that his system was "absolutely free from any possible danger from fire, even in connection with the most inflammable material." He also stated his intention to provide his company's electrical equipment to the board for inspection and approval before releasing it to the public. This was the birth of the New York Board's Bureau of Electricity. Today the Bureau of Electricity performs electrical inspections for more than 900 jurisdictions within the state of New York.

The board also maintained the Fire Patrol, the last remaining units of the Underwriters Salvage Corps left in the United States. The Fire Patrol traced its roots back to 1803 and the Mutual Assistance Bag Company, a group of New York merchants who banded together to protect their members' property. At fires, each member wore a round hat with a black brim and white crown and carried canvas bags to convey members' property and belongings to safety.

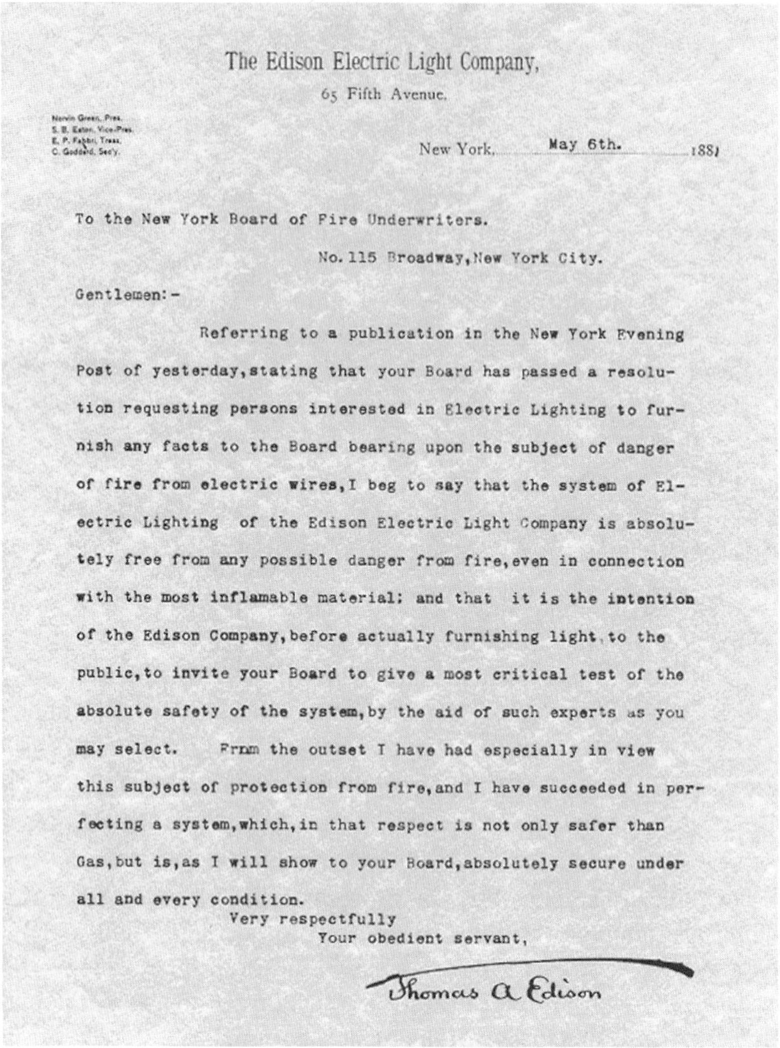

The Edison Electric Light Company,
65 Fifth Avenue,

Norvin Green, Pres.
S. B. Eaton, Vice-Pres.
E. P. Fabbri, Treas.
C. Goddard, Sec'y.

New York, May 6th. 1881

To the New York Board of Fire Underwriters.
No. 115 Broadway, New York City.

Gentlemen:-

Referring to a publication in the New York Evening Post of yesterday, stating that your Board has passed a resolution requesting persons interested in Electric Lighting to furnish any facts to the Board bearing upon the subject of danger of fire from electric wires, I beg to say that the system of Electric Lighting of the Edison Electric Light Company is absolutely free from any possible danger from fire, even in connection with the most inflamable material; and that it is the intention of the Edison Company, before actually furnishing light to the public, to invite your Board to give a most critical test of the absolute safety of the system, by the aid of such experts as you may select. From the outset I have had especially in view this subject of protection from fire, and I have succeeded in perfecting a system, which, in that respect is not only safer than Gas, but is, as I will show to your Board, absolutely secure under all and every condition.

Very respectfully
Your obedient servant,

Thomas A Edison

**FIGURE 3-3** Edison's letter to the New York Board of Fire Underwriters, claiming that his electrical system was "absolutely free" from the possibility of fire. (Courtesy of the New York Board of Fire Underwriters Collection.)

The Fire Patrol evolved from the membership of the Mutual Bag Company in 1839.[11] The Fire Patrol received alarms via the fire department telegraph system and responded with the fire department to protect property from smoke and water damage and to perform salvage operations after fires were extinguished. Patrolmen also assisted the fire department in making forcible entry. Their actions in rescuing those trapped at fires was described by Charles T. Hill in 1898 as "brilliant," and that "several of the most daring rescues" were performed by members of the Fire Patrol.[12]

On October 15, 2006, the New York Fire Patrol closed its three stations, two in Manhattan and one in Brooklyn (Figure 3-4). The Fire Patrol responded to more than 10,000

**FIGURE 3-4**
The Fire Patrol's chief mission was property protection. (Courtesy of Anthony Emanuele.)

alarms each year, protecting property and preventing or reducing fire, smoke, and water damage. The Fire Patrol also conducted fire prevention education programs. More than 90,000 children participated in its fire safety essay and poster contest each year during Fire Prevention Week. On September 11, 2001, all three Fire Patrol companies responded to the World Trade Center. Patrolman Keith Roma was among those who gave their lives operating at the incident. The insurance industry maintained salvage corps in most major cities in the United States. Salvage companies typically worked on floors below the fire, protecting property from smoke and water damage. But members are credited with numerous rescues and acts of bravery.

## FM Global, the Factory Mutual System

Stock (for-profit) insurance companies adopted fire prevention as a business strategy in the last half of the nineteenth century. The move was not merely an attempt to increase profits. The very viability of the industry was at risk from the massive conflagrations that swept through Chicago, San Francisco, Baltimore, and almost every major city. Researchers who request information regarding the "great fire" in a particular city often get the response from archivists: "Which one"?

**NOTE** Stock (for-profit) insurance companies adopted fire prevention as a business strategy in the last half of the nineteenth century.

**FIGURE 3-5**
Zachariah Allen, father of the Factory
Mutual system. (Photo courtesy of
FM Global.)

Although elected officials and much of the public were seemingly indifferent to the threat of fire, the insurance companies were not the only group that understood the value of fire prevention and protection. The textile industry was located along rivers in New England, taking full advantage of waterpower to run the mills and the large pool of workers to operate machinery. In 1822, Zachariah Allen constructed a textile mill near Providence, Rhode Island (Figure 3-5). He installed the first rotary fire pump, copper-riveted fire hose, and central heating to eliminate the need for individual stoves on the work floor. A fire wall was constructed to separate the picker room, where the hazard from fibers and lint often led to fires.[13]

Allen approached his insurance company and requested a rate reduction in consideration of his investment in fire protection and the improved risk his mill represented. The response from the insurance company board of directors was to change the insurance industry in the United States forever. "Mr. Allen, though it seems unjust, the board has decided that a fire risk is a fire risk, and we can make no reduction." Allen's reply was prophetic. "Gentlemen, the day will come, when you will regret what you admit to be unjust action on your part, good day."[14]

In 1835, the Manufacturers' Mutual Fire Insurance Company was formed. Allen was elected to the board of directors. Other factory mutuals were organized throughout New England and, in time, elsewhere in the country. The difference between the mutual system and the stock companies was significant. First, only the very best risks qualified for membership in the mutuals. Properties were inspected annually, originally by company officers. Facilities that could not pass muster had their coverage cancelled.

During the first year of existence, members of Manufacturers Mutual realized a savings of 63¼ percent of their former insurance costs.[15] As a single company however, Manufacturers Mutual was still too small to withstand the financial loss of an entire plant. Allen formed a second company, Rhode Island Mutual, and encouraged other mill proprietors to form others. They banded together as the Associated Factory Mutual Fire Insurance Companies, or the Factory Mutuals for short.

**NOTE** During the first year of existence, members of Manufacturers Mutual realized a savings of 63¼ percent of their former insurance costs.

By 1878, the company officers could no longer devote adequate time to the inspection process, and the Factory Mutuals formed an engineering unit to handle the collective inspection activities of all the companies. Originally called the Bureau of Inspections, the Factory Mutual Engineering Association soon added research, statistical analysis, and appraisal to its inspections duties. The hallmark of the Factory Mutual system was fire loss control through inspection, research, and statistical analysis.

By 1987, 42 separate mutual insurance companies had merged into three: Allendale Mutual, Arkwright Mutual, and the Protection Mutual Insurance Company. Collectively, they owned Factory Mutual Engineering and Research. In 1998, the three remaining Factory Mutuals merged to form FM Global, the twenty-fourth largest insurance organization in the United States.

FM Global attempts to reduce the risk of fire and minimize the overall financial impact if a fire does occur. FM Global Property Loss Prevention Data Sheets are engineering guidelines written to help reduce the chance of property loss caused by fire, weather conditions, and failure of electrical or mechanical equipment. The guidelines are used by industry and the U.S. Department of Defense. FM Global also develops and distributes training programs free of charge to federal, state, and local fire training organizations. Through its engineering branch, Allen's vision and Atkinson's engineering ideas remain driving forces in the field of fire prevention and protection.

## Other Mutuals

Manufacturing was not the only industry to embrace the concept of mutual insurance. Mutual insurance companies were formed within the lumber, hardware, and grain-processing industries. The Mill Mutuals were formed by grain mill and elevator operators in the Midwest. Similar to their counterparts in the textile manufacturing industry, inspection and fire prevention were key components of their risk reduction strategies.

**FIGURE 3-6**
Typical sign identifying a Mill Mutual
protected property.

The Mill Mutual Fire Prevention Bureau was established by the member grain mill
and elevator companies in 1910 to reduce the risk of fire and explosions in protected
properties[16] (Figure 3-6).

## COMMERCIAL FIRE PREVENTION AND PROTECTION PROGRAMS

The imprints of commercial fire prevention and protection organizations are on almost
every building. The smoke alarms in our homes were manufactured as a commercial enterprise. The fire prevention and protection services provided by businesses range from

design and engineering services, manufacturing, installation, testing and maintenance, and sales of all of the foregoing. Commercial entities provide fire protection and suppression services within the United States for businesses; for local, state, and federal governments; and at U.S. government installations abroad.

**NOTE** The imprints of commercial fire prevention and protection organizations are on almost every building.

## Government Contractors

The U.S. government contracts for fire prevention and protection at government installations within the United States and abroad. For the most part, fire suppression and prevention services within the United States are provided by U.S. government employees. Within the Department of Defense, active duty military personnel and civilian civil service employees are used. The military branches have developed staffing systems that are best suited for their operational needs.

Air Force bases within the United States typically have both active duty military personnel and civilian personnel serving on structural and crash firefighting companies and in support positions such as training and fire prevention. When aircraft from the base are deployed overseas, the military staff is deployed to provide fire protection and prevention. At Marine Corps installations within the United States, civilians typically provide the structural crews and support functions, and crash firefighters are Marines. The Navy and Army typically use civilian civil service firefighters for all functions.

At U.S. installations overseas and bases belonging to foreign governments where U.S. forces are deployed, fire protection and prevention services are provided in a variety of ways. Private corporations provide fire protection and prevention services at some installations, often with chief officers who are U.S. citizens and citizens of the host nation serving in the other positions. In combat zones, military firefighters provide fire prevention and protection, complete with turnout gear, weapons, and body armor.

Combat operations do not reduce the need for fire prevention and protection; they increase it. Troops deployed in the Middle East since the first Gulf War have been housed in tents. Tent cities made up of hundreds of fabric-covered structures, complete with air conditioning and electricity, have been home for our troops. The threat of fire is real, and strict policies regulating fuels, ignition sources, and adequate firebreaks must be enforced.

**NOTE** Combat operations do not reduce the need for fire prevention and protection; they increase it.

## Installation Contractors

Fire protection systems, a key element in the overall picture of fire prevention, are almost always installed by private firms. Most states require some type of professional registration or certification for designers and installers. Professional registration and certification provide a means to ensure that firms and their employees meet the minimum standards to perform the job adequately.

Issuing permits to install, remove, or modify fire protection systems provides a means of ensuring that the installation will meet minimum requirements required by code by:

- Ensuring that contractors are technically qualified through certification or registration
- Ensuring that contractors are financially qualified by requiring licensure and bonding
- Ensuring that the work meets code and is appropriate for the hazard by reviewing of a plan for the proposed work
- Providing for inspection and acceptance testing after the work is completed

Bonding and insurance are often tied to professional registration and licensure. Bonds and insurance are used to protect the public from the possibility that contractors cannot or will not adequately perform the job. To perform work, the contractor posts a bond as a guarantee of satisfactory performance.

If a contractor is hired to install fire hydrants within a jurisdiction and performs unsatisfactorily and leaves the job unfinished, the local government probably does not have the option to wait for a lawsuit to settle the issue. Unsatisfactory performance is grounds for bond forfeiture. The local government then uses the bond to pay another contractor to finish the job.

## Consulting, Maintenance, and Repair Firms

Businesses and governments often need the services of experts for short periods of time and for specific projects. The frequency of the projects and programs generally precludes the hiring of full-time employees with professional expertise, so outside consultants and firms that specialize in specific areas are retained. In the fire prevention and protection arena, the use of consultants and contractors is common (Figure 3-7). In some cases, firms even retain a consultant to help them hire the best contractor or to act in an oversight capacity to protect the interests of the firm.

Many firms that install fire protection systems also perform routine maintenance and testing. *NFPA 25: Standard for the Inspection, Testing, and Maintenance of Water-Based Fire Protection Systems* and *NFPA 72:® National Fire Alarm Code®* include specific, detailed provisions for the maintenance, inspection, and routine retesting of fire protection systems. Other standards, such as *NFPA 17: Standard for Dry Chemical Extinguishing Systems* and *NFPA 17A: Standard for Wet Chemical Extinguishing Systems*, also contain maintenance and testing requirements.

*NFPA 25, Standard for the Inspection, Testing, and Maintenance of Water-Based Fire Protection Systems*, emphasizes the importance of maintenance and testing:

> History has shown that the performance reliability of a water-based fire protection system under fire-related conditions increases where comprehensive inspection, testing, and maintenance procedures are enforced. Diligence during an inspection is important.[17]

**FIGURE 3-7**

Maintenance and testing by a fire protection contractor. (Courtesy of Tom Herman, Eagle Fire Protection.)

In addition to retaining the services of consultants, some firms specifically hire maintenance and testing contractors other than those that installed the systems. Whether the practice results in better service and an increased level of protection is wholly dependent on the quality and integrity of the contractors. If a particular contractor cannot be trusted to perform routine maintenance and inspection, how much confidence should there be in the system the contractor installed?

**NOTE** If a particular contractor cannot be trusted to perform routine maintenance and inspection, how much confidence should there be in the system the contractor installed?

Inspection by municipal fire prevention officials should not be considered absolute protection against contractor fraud or poor work. Required inspections and acceptance tests indicate that systems comply with minimum code provisions and performed acceptably in the required tests (Figure 3-8).

## Consulting Engineers

In addition to working with fire protection systems, licensed fire protection engineers are often retained to perform fire protection and life safety surveys, develop and review fire

**FIGURE 3-8**
Tags and test reports
are evidence of required
maintenance and testing.
(Courtesy of Tom Herman, Eagle
Fire Protection.)

safety and fire evacuation plans, and develop strategies for solving specific fire protection
and fire prevention issues. They may assist in the establishment and training of industrial
fire brigades, perform fire investigation and fire loss surveys, and act as expert witnesses
in court cases.

## Third-Party Inspection and Certification

The model building and fire codes contain provisions for code officials to accept reports
from qualified individuals and firms as evidence of compliance with code:

> The code official is authorized to conduct such inspections as are deemed necessary
> to determine the extent of compliance with the provisions of this code and to approve
> reports of inspection by approved agencies or individuals.[18]

The resources required to staff, train, and equip an inspections bureau are beyond
the means of some local governments. Others make the conscious decision to retain the
services of a **third-party inspection** agency. Some fire service professionals express concern that the "fox is watching the henhouse," but adequate safeguards can be instituted

to ensure fair, competent inspection and approval. This is not self-inspection by the contractor, and in most jurisdictions, companies that install systems cannot inspect them. This policy ensures that collusion between contractors to approve each other's work is prevented.

> **third-party inspection** inspection or testing by an approved nongovernmental agency that is forwarded to the code official for action or an inspection performed on behalf of the local government.

Many third-party engineering and inspection agencies limit themselves to inspection work and neither design nor assist in the installation process. Many state building codes permit local governments to contract with third-party agencies in lieu of employing plan review and inspections personnel. This is particularly popular in small jurisdictions that cannot justify the cost of full-time personnel with the necessary certifications and experience. In most cases, the municipal government is still responsible for administering the code and ensuring that the third-party agency is performing its duties in accordance with the code.

In any case, the fact that a fire official chooses to accept reports from third-party agencies does not relieve the fire official of the responsibility for approval. The fire official approves or rejects the work based on the report from the third party. The fire official still bears the legal responsibility for the approval, which cannot be delegated without a specific legal enablement.

**NOTE** The fact that a fire official chooses to accept reports from third-party agencies does not relieve the fire official of the responsibility for approval.

# PRIVATE ASSOCIATIONS AND NOT-FOR-PROFIT ORGANIZATIONS

Private and not-for-profit organizations, created to provide a service to the general public, play a key role in the fire prevention programs of governments and private industry. Most of the codes, standards, and recommended practices used in building regulation, fire protection, and fire prevention are produced through the efforts of private associations and not-for-profit organizations.

## Underwriters Laboratories

Research into the origins of the major organizations involved in fire prevention and fire safety in the United States leads back to small groups of people who shared a common interest in reducing the amount of fire waste in a young republic that called itself the United States of America. The strategies that they developed and the plans that were implemented often proved successful beyond the wildest dreams of most nineteenth century Americans. UL is just such an example.

In 1893, a young engineer named William H. Merrill was hired by the Chicago Underwriters Association. There were problems with the automatic fire alarm systems in the city, and the World's Fair was coming to Chicago. Installations and displays at the fair featured a new technology, called electricity, that had the insurance underwriters worried. Merrill had come from Boston and had suggested to the Boston Board of Fire Underwriters that a testing laboratory be established there, but his idea was rejected.

The Chicago Underwriters gave Merrill his laboratory—a workbench on the third floor of Fire Insurance Patrol Station #1. The lavish sum of $350 was appropriated for equipment and some chairs and to hire a clerk and a helper for Merrill. In his room above the fire patrol horses, Merrill tested electrical equipment destined for the World's Fair, ensuring its safety. He was so successful that the Underwriters Electrical Bureau was born, soon to draw the attention of the National Board of Fire Underwriters. It funded the program, which became the Electrical Bureau of the National Board.[19]

By 1903, the bureau had become so large that it was incorporated as UL. UL filled a void that existed in fire prevention and protection as well as other areas of safety. There were few standards for the new technologies and products that were rapidly emerging. Kerosene, sold as a liniment under the name American Medicinal Oil, in 1829 was later found to be a substitute for whale oil.[20] Kerosene was later used as fuel for heaters and stoves. Fires resulted from improper storage, poorly designed and constructed appliances, and the simple lack of knowledge on the part of an unsuspecting public. In *Symbol of Safety*, a history of the early years of UL, Harry Chase Brearley described the hazards created by progress: "One evidence of this is found in fire losses which, in the United States, increased more than one-thousand percent between 1865 and 1922, while population increased but two-hundred-percent."[21]

UL was involved with some of the earliest fire tests of building materials and assemblies that were conducted in the United States. Fire tests were jointly conducted by the National Board of Fire Underwriters, NFPA, Associated Factory Mutual Engineering, and Bureau of Standards at UL from 1912 to 1917.[22] These tests were among the earliest well-documented scientific fire tests performed in the United States and were the basis for the model building codes.

UL tests and evaluates products at UL test facilities worldwide and develops standards. Twenty-four of the more than 800 UL standards were referenced in the 2000 International Building Code. In 2002, UL marks appeared on 17 billion products. UL, an organization that began in Chicago Fire Patrol Station #1, has always been closely affiliated with the NFPA, Factory Mutual Engineering, National Board of Fire Underwriters, and Bureau of Standards, organizations that were and still are at the forefront of fire prevention.

## Codes and Standards Organizations

In Chapter 4, "Fire Prevention Through the Codes Process," the model code organizations—the NFPA, International Code Council, and others—are discussed at length. In this chapter, only the **standards**-making organizations not included in Chapter 4 are discussed.

> **standard**  a rule for measuring or a model to be followed; how to do something and what materials to use; also known as referenced standard.

#9

## American Society for Testing and Materials

Organized in 1898, the American Society for Testing and Materials (ASTM) is one of the largest voluntary standards-developing organizations in the world. ASTM's technical documents are used in manufacturing, management, procurement, codes, and regulations. The 11,000 standards address a broad range of subjects and materials from steel, petroleum, and medical devices to property management and consumer items. More than 200 ASTM standards are referenced in the International Building Code and International Fire Codes; those listed in Figure 3-9 are among those that address fire protection.

#9

## American National Standards Institute

The American National Standards Institute (ANSI) is a 90-year-old organization dedicated to coordinating the efforts of U.S. standards-making organizations. The ANSI does not develop standards. In 1916, American standards-writing organizations banded together to avoid duplication, waste, and conflicting standards. In response to the unification of European markets and standards and the North American Free Trade Agreement (NAFTA), the ANSI's international presence was expanded. The ANSI is

| | |
|---|---|
| ASTM E-84 | Standard Test Methods for Surface Burning Characteristics of Building Materials |
| ASTM E-108 | Standard Test Methods for Fire Tests of Roof Coverings |
| ASTM E-119 | Standard Test Methods for Fire Tests of Building Construction Materials |
| ASTM E-136 | Standard Test Methods for Behavior of Materials in a Vertical Tube Furnace at 750° |
| ASTM E-605-93 | Standard Test Methods for Thickness and Density of Sprayed Fire-resistive Materials Applied to Structural Members |
| ASTM E-736 | Standard Test Methods for Cohesion/Adhesion of Sprayed Fire-resistive Materials Applied to Structural Members |
| ASTM E-814 | Standard Test Methods of Fire Tests of Through-penetration Fire Stops |

**FIGURE 3-9**
ASTM standards for fire protection.

the U.S. representative to the International Organization for Standardization (ISO), and the International Electrotechnical Commission (IEC).

## American Institute of Architects

A powerful force within the code development process, the American Institute of Architects (AIA) is a 70,000-member association that represents architects. Approximately 82 percent of the members are licensed through state registration systems.[23] Architects design the buildings and structures in which we live, work, shop, worship, go to school, and recreate. Through the practice of their profession and through their advocacy in the code process, the organization has a huge impact on the fire safety of every American.

## Society of Fire Protection Engineers

Established in 1950, the Society of Fire Protection Engineers (SFPE) is a professional society representing engineers practicing in the field of fire protection. The society has approximately 3,500 members in the United States and abroad. The mission of the organization is to advance the science and practice of fire protection engineering and its allied fields, to maintain high ethical standards among its members, and to foster fire protection engineering education. The SFPE supports and presents fire protection seminars and conferences, develops and disseminates technical information on fire protection engineering, and assists in the licensing examination process for professional engineers.

## National Association of State Fire Marshals

A relatively young organization, the National Association of State Fire Marshals (NASFM) was organized in 1990. The NASFM represents the most senior fire official of each of the fifty states and the District of Columbia. Although the responsibilities of state fire marshals vary from state to state, most fire marshals are responsible for code adoption and enforcement, fire and arson investigation, fire incident data reporting and analysis, and public education. They also serve as fire safety consultants for governors and state legislatures. Most state fire marshals are appointed by the governor or another high-ranking state official such as the insurance commissioner. The NASFM's mission is expressed in two strategic goals:

1. To protect human life, property, and the environment from fire
2. To improve the efficiency and effectiveness of state fire marshals' operations

## TRADE ASSOCIATIONS

Trade associations representing manufacturers, installers, and fire protection equipment manufacturers play an important role in the field of fire prevention. These include the American Concrete Institute, American Forest and Paper Association, Gypsum Association,

#10

American Iron and Steel Institute, Portland Cement Association, Brick Institute, American National Fire Sprinkler Association, National Fire Sprinkler Association, and Builders Hardware Manufacturers' Association.

Associations develop standards, design and test **fire resistance–rated assemblies**, maintain materials testing laboratories, provide technical information and training about their products, and represent their members in the codes and standards development process. Trade association directories and design manuals are used by industry, design professionals, and regulators.

| fire resistance–rated assemblies | assemblies of materials designed and tested to retard the passage of heat, smoke, and fire for a given period of time. |
| --- | --- |

## SUMMARY

Reducing or eliminating the threat of fire has been a goal of government and industry since the earliest days of the American colonies. Federal, state, and local governments have programs aimed at protecting the public and reducing fire deaths and injuries. Public fire prevention programs literally could not exist as we know them without the products and efforts of private sector fire prevention organizations.

Private fire prevention and fire protection programs generally fall into three categories: those undertaken by business as part of a risk management system, those that provide fire prevention and protection as a profit-making business service, and those that are not-for-profit entities operating in the public interest. Often the roles, products, and missions of the organizations come together and are complementary. Some of the earliest fire prevention programs in the United States were organized by the owners of mills and factories as methods of reducing both the threat of fire and the high cost of insurance.

The early fire prevention efforts of the insurance industry evolved into our current system of construction codes and regulations and municipal fire protection. Business and industry in the United States continue to manage fire risk in order to enhance profits and remain viable. Many companies are in the business of providing services aimed at preventing or reducing the threat of fire. Design professionals, installation and maintenance contractors, and third-party inspection agencies are key players in fire prevention.

The bulk of fire prevention and protection codes, standards, and recommended practices are developed and maintained by not-for-profit organizations and associations whose sole purpose is the protection of life and the reduction of property loss through fire and other catastrophic events.

## REVIEW QUESTIONS

PG 50,59,63 1. List three categories of organizations involved in fire prevention and protection.

PG 49 2. In private industrial facilities, are fire safety and plant security both mandated by law?

PG 51 3. List the three categories used by the ISO to determine a community's PPC.

PG 50&53 4. What system was developed by the insurance industry in response to the losses resulting from Hurricane Andrew and the Northridge Earthquake?

PG 57 5. What is the name of the organization that develops Property Loss Prevention Data Sheets?

PG 60 6. Permits required to install fire protection systems are intended to ensure what four items?

PG 60 7. What NFPA standard covers the maintenance of automatic sprinklers?

PG 60 8. What is the purpose of contractors' bonds?

PG 61 9. List three organizations that develop standards.

PG 66 10. List four trade associations involved in fire prevention and protection.

## DISCUSSION QUESTIONS

1. You receive a call from the owner of a large industrial facility that was inspected by your office. The facility has been well managed, and there are no outstanding violations. The manager requests your assistance. An insurance engineer has inspected his building and has an extensive list of demands. You are requested to call the insurance engineer and inform him that because the facility complies with code, the manager is not obligated to comply. What is your response to the manager?

2. Using the same scenario, what if the insurance engineer orders a condition that conflicts with your adopted fire code?

## CHAPTER PROJECT

Select one of the private fire prevention organizations discussed in the chapter and describe specific ways in which their fire prevention and protection efforts directly affect you and your family. Your examples should be specific and identify the financial (or other) impact as it relates to your daily lives.

## ADDITIONAL RESOURCES

In-depth information on many of the subjects discussed in this chapter can be found in the following texts and publications and at these Web sites.

American Institute of Architects at www.aia.org.

American Insurance Association at www.aiadc.org.

American National Standards Institute at www.ansi.org.
American Society for Testing and Materials at www.astm.org.
FM Global (Factory Mutual Engineering) at www.fmglobal.com.
Insurance Services Office Public Protection Classification program at www.isomitigation.com.
National Association of State Fire Marshals at www.firemarshals.org.
New York Board of Fire Underwriters at www.nybfu.org.
Society of Fire Protection Engineers at www.sfpe.org.
Underwriters Laboratories at www.ul.com.

## NOTES

1. A.L. Todd, *A Spark Ignited in Portland* (New York: McGraw-Hill, 1966) page 26.
2. *Fifty Years of Civilizing Force* (New York: Frederick A. Stokes, 1916), page 46.
3. John L. Bryan, *Automatic Sprinklers and Standpipes* (Quincy, MA: National Fire Protection Association, 1990), page 60.
4. Peter J. McKeon, *Fire Prevention* (New York: The Chief Publishing, 1912), page 6.
5. Kevin Cassidy, *Fire Safety and Loss Prevention* (Boston: Butterworth-Heinemann, 1992), page 8.
6. National Fire Data Center, *Technical Report 057 Chicken Processing Plant Fires, Hamlet North Carolina and North Little Rock Arkansas.*
7. *ISOs PPC Program* (New York: ISO Properties, Inc., 2001), page 2.
8. Ibid., page 3.
9. *Evaluating Building Code Effectiveness* (New York: Insurance Services Office, Inc., 2000), page 5.
10. Ibid., page 6.
11. Charles T. Hill, *Fighting a Fire* (New York: The Century Co, 1898) page 227.
12. Ibid., page 230.
13. *Prevention, A Factual History of Property Loss Prevention and Control, Including the Role Played by Factory Mutual*, Factory Mutual Engineering, 1998, page 10.
14. *The Factory Mutuals* (Providence: Manufacturers Mutual Fire Insurance Company, 1935), page 33.
15. Ibid., page 43.
16. John Bainbridge, *Biography of an Idea* (New York: Doubleday, 1952), page 217.
17. *NFPA 25: Standard for the Inspection, Testing, and Maintenance of Water-Based Fire Protection Systems*, 2002 ed. (Quincy, MA: National Fire Protection Association, 2002) A.1.2.
18. 2009 *International Fire Code* (Falls Church, VA: International Code Council, 2009), page 11.
19. Harry Chase Brearley, *A Symbol of Safety* (Garden City, NY: Doubleday, Page and Company, 1923) page 18.
20. National Board of Fire Underwriters, *Safeguarding the Home Against Fire*, 1918, page 49.
21. Brearley, page 6.
22. Associated Factory Mutual Insurance Companies, *Fire Tests of Building Columns*, 1917, page 17.
23. American Institute of Architects, *AIA Firm Survey 2000–2002*, Washington, DC.

# 4

# Fire Prevention Through the Codes Process

## LEARNING OBJECTIVES

Upon completion of this chapter, you should be able to:

- Describe the origin of the model code system in the United States.

- List the major model code organizations and describe the evolution of model code organizations in the United States.

- Describe the code change process used by the model code organizations.

- Describe the methods of code adoption by states and local governments.

- Discuss the impact of the agendas of groups participating in the model code process.

# MODEL CODES

The fire prevention and code enforcement efforts of state and local governments are driven by codes. **Codes** are merely systematically arranged bodies of laws or rules. Codes tell us what to do or what not to do. Examples are the *United States Code*, *Code of Virginia*, and *Code of the County of Fairfax*. These are the codified laws of the United States, Commonwealth of Virginia, and Fairfax County, Virginia, respectively. The term *codified* refers to the fact that the laws are arranged according to a system.

| **code** | a systematically arranged body of rules; when and where to do or not to do something. |

**NOTE** Codes tell us what to do or what not to do.

Model codes are technical rules that are developed by organizations and are made available for governments to formally accept and put into effect within the jurisdiction. We refer to this formal acceptance as **adoption** of the code. Code adoption is discussed in depth later in this chapter. Model codes are available for governments to adopt free of charge. Why would organizations invest thousands of hours of hard work only to offer their products to the public (through their governments) free of charge? Why would a state or local government decide to adopt and make the rules developed by an outside special interest group the law in its jurisdiction? To truly appreciate the complex codes system in use in our nation, a look at the fire prevention and protection failures of the past is necessary.

| **adoption** | the acceptance and formal approval of a model code by the governing body of a jurisdiction. |

## The Development of Model Codes

Chapter 1 discusses some of the significant fires that plagued our country from colonial days. In the aftermath of every major fire, there was an outcry for action followed by the passage of either a law or laws. These laws were overlooked or forgotten as soon as compliance became inconvenient or inexpedient. The political process is not necessarily the best medium for development of highly technical rules. Perhaps the most striking example was the city of Chicago, where within 3 years of the great fire that killed 300 and destroyed more than 17,000 buildings, fire safety conditions were identified by insurance inspectors to be worse than before the fire![1] If governments were not capable of technical code development, then who was?

The fire insurance industry failed in its attempts to self-regulate insurance rates and commissions. The idea was to get the entire industry to agree to charge uniform rates and pay their agents standard commissions as a method of ensuring they could maintain adequate reserves to pay catastrophic claims. Greed won out, leaving them in a no-win situation. Inept or unscrupulous companies offered insurance at rock-bottom prices. The thin profit margins left little reserves to pay claims after massive fires. When disaster struck, the cut-rate companies simply declared bankruptcy, leaving their policyholders with unpaid claims. The legitimate companies were then left to feel the wrath of politicians who crafted punitive laws in response to the public outcry against the bankrupt companies.

In self-defense, the insurance industry embarked on another method of ensuring profits—the prevention of fires and the reduction of conflagration potential through the development of codes and standards. The National Board of Fire Underwriters' (NBFU) first attempts at developing standards had involved setting specifications for "first class" woolen mills and sugar houses, which would qualify for reduced rates by virtue of their class rating, and for a whale oil substitute sold under the name of "kerosene oil."[2] In 1892, D.W.C. Skilton, president of the NBFU (Figure 4-1), set the tone for the industry's fire prevention efforts for the better part of the next century:

> The old theory . . . that risk should be written as found, and a rate adequate to the hazard be charged is fast becoming obsolete, and today all local and district associations, and all syndicates for writing great industries are aiming to secure improvements in construction and greater care, and all favor the introduction of automatic and other appliances for the prevention and extinguishing of fires, the inducement to the assured being a greatly reduced rate for this lessening of hazard.[3]

**FIGURE 4-1**

D. W. C. Skilton called for the insurance industry to reduce its risks by improving the fire safety conditions of its protected properties. (Source: Harry Chase Brearley, *Fifty Years of Civilizing Force*, Frederick A. Stokes, New York, 1916, p. 53.)

In the same year, a marked increase in fires at facilities considered to be better risks underscored a new hazard associated with economic success and innovation—electricity. An emergency meeting held in New York resulted in the formation of the NBFU's Underwriter's International Electrical Association and development of the *National Board Electrical Code*, known today as the *National Electric Code®* (*NEC®*; NFPA 70). By 1901, the code was being enforced by 125 municipal governments. The NBFU transferred responsibility of the code to the National Fire Protection Association (NFPA)in 1911. The *NEC®* may well be the most widely used model code in the world.

A model building law for the state of New York was introduced to the state legislature in the early 1890s, but it was defeated by the efforts of representatives of Buffalo, Jamestown, and other small cities. By 1896, the NBFU voted to expand the proposed building code, to be called the *National Board's Model Building Law*. In 1905, the first edition of the Board's *National Building Code* was distributed free of charge to all cities with a population more than 5,000 and to contractors, architects, fire marshals, and technical schools (Figure 4-2).[4] The NBFU later developed the *National Fire Prevention Code*. The NBFU published the *National Building Code* and *National Fire Prevention Code* through 1976.

**FIGURE 4-2**
The 1905 National Building Code, first model building code in the United States. The National Building Code is the basis for model building codes in use today.

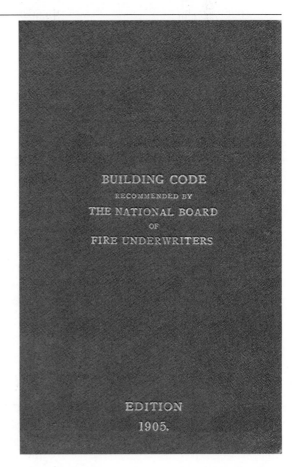

## MODEL CODE ORGANIZATIONS OF THE TWENTIETH CENTURY

Although several large cities had their own locally developed building and fire codes, many jurisdictions adopted the NBFU codes. Other organizations emerged in the twentieth century that also developed model codes and standards that accompanied the model codes. A standard is a guide or rule to be followed. Codes tell us what to do. Standards tell us how. The building codes require hospitals to be equipped with automatic sprinklers and refer to a sprinkler standard for specifics on design, materials, installation, and performance.

The twentieth century ended with a drastically different lineup in the model code development process. The NBFU was out of the code business and had been absorbed as a part of Insurance Services Organization (ISO). The NFPA had grown from a handful of New England insurance engineers into a 75,000-member international organization. The NFPA's Web site, www.nfpa.org, lists the organization's mission: "To reduce the worldwide burden of fire and other hazards on the quality of life by developing and advocating scientifically based consensus and standards, research, training and education."

The nation's three largest model code organizations had abandoned their own regionally developed codes and had cooperatively developed a single set of construction and safety codes. Their success at working together would pave the way to consolidation into the International Code Council (ICC), a 50,000-member organization of construction, safety, fire and other public officials, design professionals, installers, manufacturers, and builders. The events that led to the current model code system and the players in the model code process shed considerable light on what to expect in the future.

## THE REGIONAL MODEL CODE ORGANIZATIONS

The system of regional codes that gradually evolved, supplanting the NBFU, began in the 1920s with the development of a building code on the West Coast by the Pacific Building Officials Conference. By 1950, there were three regional code groups. Their success led the NBFU to abandon code development altogether and focus attention on other issues. Three major regional codes evolved in the twentieth century, and almost every state adopted at least one of them. The differences were not always subtle, leading in some cases to increased costs for design and construction. By the late 1980s, the national map of adopted codes resembled a puzzle. Figure 4-3 is based on research by the Council of American Building Officials in 1998 and shows the influence of the regional code system.

### Building Officials and Code Administrators, International

Building Officials and Code Administrators, International (BOCA) was established in 1915 by building commissioners from nine northeastern states and Canada as the Building Officials Conference of America to "discuss the principles underlying ordinances

## FIGURE 4-3

Model codes used throughout the United States in 1998.

| BOCA/ NATIONAL CODES | ICBO/ UNIFORM CODES | SBCCI/ STANDARDS CODES | TWO OR MORE CODES | STATE-DEVELOPED CODES |
|---|---|---|---|---|
| Connecticut | Alaska | Alabama | Kansas | New York |
| Delaware | Arizona | Arkansas | Michigan | Wisconsin |
| Illinois | California | Florida | Missouri | |
| Kentucky | Colorado | Georgia | Oklahoma | |
| Maine | Hawaii | Louisiana | Texas | |
| Maryland | Idaho | Mississippi | | |
| Massachusetts | Indiana | North Carolina | | |
| New Hampshire | Iowa | South Carolina | | |
| New Jersey | Minnesota | Tennessee | | |
| Ohio | Montana | | | |
| Pennsylvania | Nebraska | | | |
| Rhode Island | Nevada | | | |
| Vermont | New Mexico | | | |
| Virginia | North Dakota | | | |
| West Virginia | Oregon | | | |
| | South Dakota | | | |
| | Utah | | | |
| | Washington | | | |
| | Wyoming | | | |

related to building."[5] In 1950, BOCA published its first *Basic Building Code*. The organization maintained building, mechanical, fire prevention, plumbing, and property maintenance codes through 1999. Before consolidation with the other ICC members, BOCA served the northeast, mid-Atlantic, and midwestern states.

## Southern Building Code Congress International

The Southern Building Code Congress International (SBCCI) was established in 1940 and published the first edition of the *Standard Building Code* in 1945. The SBCCI also published plumbing, mechanical, property maintenance, outdoor sign, and building conservation codes through 1999. The *Standard Fire Code* was cooperatively developed with the Southeastern and the Southwestern Fire Chief's Associations under a memorandum of understanding. Final voting on the *Standard Fire Code* was limited to fire service representatives. Before consolidation with the other ICC members, the SBCCI served the southeastern and south central states.

## International Conference of Building Officials

The International Conference of Building Officials (ICBO) was established in 1921 as the Pacific Building Officials Conference, and it published the first edition of its *Uniform Building Code* in 1927.[6] The ICBO went on to publish numerous related codes, including those for property maintenance, outdoor signs, building security, and building conservation. The ICBO published the *Uniform Plumbing Code* and *Uniform Mechanical Code* in cooperation with the International Association of Plumbing and Mechanical Officials (IAPMO) and the *Uniform Fire Code* in cooperation with the Western Fire Chiefs Association (WFCA). The IAPMO and WFCA retained copyrights for their respective codes, although the codes were published and distributed by ICBO. Before consolidation with the other ICC members, the ICBO served the western states, Alaska, and Hawaii.

The Uniform Building Code (UBC) was incorporated into the Department of Defense Military Handbook 1008, *Fire Protection for Facilities Engineering, Design, and Construction*, for general building code requirements. The UBC was superseded by the implementation of *Unified Facilities Criteria* (UFC 1-200-01) on July 31, 2002, which incorporated the *International Building Code*.

## International Code Council (ICC)

The ICC was established in 1994 as an umbrella organization consisting of representatives of BOCA, ICBO, and SBCCI for the express purpose of developing a single set of model codes for the United States. Calls for a single set of codes had been heard from various groups over the years, but world events in the last decade of the twentieth century put the regional code concept under the spotlight and created an atmosphere for cooperation among the regional code organizations.

**NOTE** The ICC was established in 1994 as an umbrella organization consisting of representatives of BOCA, ICBO, and SBCCI for the express purpose of developing a single set of model codes for the United States.

By the early 1990s, the European Union had grown to 15 countries, and as part of its effort to form a large common market for goods and services, it had established a Standing Committee on Construction and had harmonized standards for construction and construction materials. The North American Free Trade Agreement (NAFTA), linking Canada, the United States, and Mexico, was eliminating trade barriers on the North

American continent. Was our regional code system, with its technical disparities, a threat to American competitiveness? The members of the three model code groups did not wait for an answer to come from Congress or some think tank. They formed the ICC and began work on the "I" Codes. The *International Plumbing, Mechanical,* and *Private Sewage Disposal Codes* were published in 1996. The *International Building, Fire,* and *Property Maintenance Codes* were published in 2000.

The success in developing a single set of codes naturally led to considering a second step. If the three model code groups could develop one set of codes, why bother having three model code groups? On January 21, 2003, as a result of overwhelming support by the membership of all three groups, BOCA, ICBO, and SBCCI were consolidated, merging their memberships, assets, and staffs. The consolidation created an association of more than 50,000 members, whose codes are in use in 45 states and thousands of jurisdictions.[7]

The ICC publishes 14 model codes as shown in Figure 4-4.

**FIGURE 4-4**

Model codes developed and maintained by the International Code Council.

| CODE | TOPIC/AREA |
|---|---|
| *International Building Code* | Commercial structures, multifamily residential |
| *International Energy Conservation Code* | Energy conservation for mechanical, lighting systems |
| *International Existing Building Code* | Improving and upgrading existing structures to conserve history and resources |
| *International Fire Code* | Fire prevention and protection |
| *International Fuel Gas Code* | Gas-fueled systems and appliances |
| *International Mechanical Code* | Mechanical systems |
| *ICC Performance Code* | |
| *International Plumbing Code* | Plumbing in commercial and multifamily residential |
| *International Private Sewage Disposal Code* | Design, installation, and maintenance for private sewage disposal systems |
| *International Property Maintenance Code* | Property maintenance |
| *International Residential Code* | One- and two-family dwellings |
| *International Zoning Code* | Zoning |
| *ICC Electrical Code* | Administrative provisions for enforcement of the *National Electrical Code*® |
| *International Urban-Wildland Interface Code* | Fire protection requirements for buildings constructed near wildland areas |

## International Fire Code Council

The International Fire Code Council (IFCC) was established by the ICC to represent the common interests of the fire service and the ICC by providing leadership and direction on matters of fire and life safety and to meet government, industry, and public needs. The IFCC was formed before the development of the *International Fire Code* (IFC). Fire service influence on the development of the *IFC* through the IFCC is readily apparent. The IFC is a comprehensive, user-friendly document, clearly designed with inspectors in the field in mind.

**NOTE** The IFCC was established by the ICC to represent the common interests of the fire service and the ICC.

The IFCC is composed of 15 members, one each from the eight International Association of Fire Chiefs regions, four representatives from the National Association of State Fire Marshals, and three at-large members. Through the efforts of the IFCC, memorandums of understanding (MOUs) have established set-aside positions on code development committees for fire service members. Never has the fire service had greater influence in the model code process on fire-related matters.

## THE NATIONAL FIRE PROTECTION ASSOCIATION

There were very few comprehensive building regulations in effect in the 1800s. One of the significant factors affecting an owner's decision to consider fire-safe design and construction for a structure or facility was fire insurance. New England was the home of many of the nation's cotton and woolen mills because of the availability of moving water to power the machinery and equipment. Fire insurance companies that offered reduced premiums for sprinklered buildings or simply required the installation of fire sprinklers often had requirements for installation developed by their engineers.

Nine different standards for sprinkler pipe size and head spacing were being used by insurance companies within 100 miles of Boston in the late 1800s.[8] In an effort to develop a standard that was acceptable by all the companies, fire underwriters formed an association to work toward uniformity in 1896. The group adopted the name of National Fire Protection Association.

By 1904, the NFPA's active membership included 38 stock fire insurance boards and 417 individuals, most of whom were from the insurance industry. The first fire department officer to join the NFPA was Battalion Chief W.T. Beggin of the Fire Department City of New York (FDNY), who became a member in 1905, the year the NBFU first published the *National Building Code*. H.D. Davis, Ohio State Fire Marshal, joined the same year. In 1911, the NFPA assumed maintenance of the *National Electrical Code*® from the NBFU Electrical Committee and has published it ever since. The NFPA was incorporated in 1930.

The NFPA publishes almost 300 codes, standards, and recommended practices developed by more than 205 technical committees. Many of the documents originated as NBFU pamphlets and carried the subtitle "as recommended by the National Fire Protection Association." The NFPA's recommended practices are one of the great, untapped resources for municipal fire officials. They are not designed to be enforceable documents, but they are a guide to good practice within given industries. Alerting business owners to the existence of these documents may do more for the prevention of fires than simply enforcing the code.

**NOTE** The NFPA publishes almost 300 codes, standards, and recommended practices developed by more than 205 technical committees.

**NOTE** The NFPA's recommended practices are one of the great, untapped resources for municipal fire officials.

The NFPA's *NEC®* is perhaps the most widely used code in the United States. It was the electric code referenced by all three of the model codes groups, and it has been incorporated by reference into the *International Building Code. NFPA 101®: Life Safety Code®* addresses occupant safety in buildings with regard to the establishment and maintenance of exit facilities. It is neither a building nor a fire code but addresses some features of each. It has requirements for sprinklers, fire alarm systems, and rated construction as elements of or protection for the means of egress. It mandates employee training and drills in certain occupancies. Compliance with the *Life Safety Code®* is required for health care facilities as a condition of federal health insurance.

The *Life Safety Code®* grew out of the Triangle Shirtwaist fire, which occurred in a New York garment factory in 1911 (Figure 4-5). Until then, the NFPA had limited its efforts to engineering aimed at property protection. As an arm of the fire insurance industry, the protection of property was its stated mission. Among the public that watched the horror unfold outside the Asch Building, where the Triangle Shirtwaist Factory was located, was a woman named Frances Perkins. She watched as young women jumped from the upper floors of the 10-story loft building to the sidewalk below. There were 147 fatalities, most of them young women. Many were immigrants, grateful to have a job.

Few codes for construction or worker safety existed. The *New York City Building Code* required exit doors to swing into stair landings in the direction of egress "if practicable."[9] Because the stairs had no landings, it was not "practicable" and the doors swung against the flow of egress. In 1913, Frances Perkins, who had witnessed the Triangle disaster, was a guest speaker at the NFPA's annual meeting. She called for NFPA to take up the cause of life safety from fires. (Later appointed Secretary of Labor by President Franklin Roosevelt, she was the first woman appointed to a presidential cabinet.) By 1916, the NFPA had established a committee on the safety for life from fire and published a pamphlet on the use of outside stairs for fire exits. In 1922, the *Building Exits Code* was published, later to be renamed the *Life Safety Code®*.

In 2002, the NFPA published *NFPA 5000:™ Building Construction and Safety Code.™* The code was developed as an alternative to the ICC's I Codes after a series of disputes

**FIGURE 4-5**

The Triangle Shirtwaist fire led to the development of the *Life Safety Code.*® (Source: Joseph Kendall Frietag, *Fire Prevention and Protection,* Wiley, New York, 1912, p. 187.)

with ICC's legacy organizations. The NFPA partnered with the American Association of Plumbing and Mechanical Officials; the American Society of Heating, Refrigerating and Air-Conditioning Engineers (ASHRAE); and the WFCA to offer a set of companion codes. *NFPA 5000*™: *Uniform Plumbing, Uniform Mechanical,* and *Uniform Fire Codes* are distributed by the NFPA as the "C3" Codes. These efforts were opposed by The American Institute of Architects (AIA), Building Owners and Managers Association (BOMA), National Association of Home Builders (NAHB), and others, fearing conflicting code provisions in different states and cities. NFPA 5000 was initially adopted by the State of California, even though no other state or major city had adopted it. In 2003, California adopted the International Building Code.

## CODE CHANGES

With the consolidation of the three regional model code groups into the ICC, there are only two major groups involved in the process, the ICC and the C3 group, comprising the NFPA, ASHRAE, IAPMO, and WFCA. Both groups use a consensus process to develop and maintain their documents, but both disagree on what the term *consensus* really means. In general usage, it means "an opinion held by all or most," or "general agreement."[10] The term has been batted around and used both as a marketing tool and a means to stifle competition.

## The International Code Council Code Change Process

The ICC code change process occurs in an 18-month cycle. Any interested person may submit a code change by simply downloading the code change form or may request a form from the ICC. The process involves seven steps as shown in Figure 4-6. Committees at hearings are made up of a cross-section of members, including code officials, design professionals, installers, and manufacturers. What sets the ICC apart from the NFPA process in which all members may vote is that only code officials representing member jurisdictions are authorized to vote in the ICC process. The number of voting delegates is based on the population of the jurisdiction.

**ICC Code Development Process**

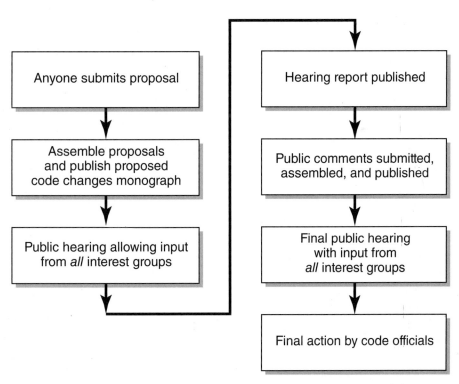

**FIGURE 4-6**
The ICC code development process.

**FIGURE 4-7**

Under the ICC system, the number of voting delegates is determined by the population of the jurisdiction.

| NUMBER OF CITIZENS SERVED | NUMBER OF VOTING DELEGATES |
|---|---|
| Up to 50,000 | 4 |
| 50,001–150,000 | 8 |
| Over 150,000 | 12 |

**NOTE** What sets the ICC apart from the NFPA process in which all members may vote is that only code officials representing member jurisdictions are authorized to vote.

The number of voting delegates shown in Figure 4-7 is applicable for each agency that enforces one or more codes within the jurisdiction. If a city of 100,000 has a building department that enforces the *International Building Code* and a fire department that enforces the *International Fire Code*, both organizations may send eight voting delegates to the code hearings, for a total of 16 delegates. It is important to note that to be a voting delegate, the employee must be directly involved in the administration, formulation, or enforcement of the code.

This system gives the fire service a significant voice in the code development process. This is the opportunity that many in the fire service have worked toward and waited for for years—and warrants significant effort in the future. The old adage "Be careful what you pray for; you might get it," rings true with the establishment of the ICC. The fire service is truly an equal partner in the code development process. Lack of fire service influence in the process can no longer be an excuse for building code provisions that the fire service finds unacceptable. Design professionals and developers will rightly ask, "Where were you guys?"

**NOTE** Lack of fire service influence in the process can no longer be an excuse for building code provisions that the fire service finds unacceptable. Design professionals and developers will rightly ask, "Where were you guys?"

## The NFPA Code Change Process

In the NFPA process, every member is assigned to one or more membership categories as shown in Figure 4-8. NFPA committees are structured so that not more than one-third of any committee represents a single interest group. The NFPA's code change process requires 104 weeks and relies on meetings of the 205 individual technical committees.

**FIGURE 4-8**
NFPA uses a system of nine member categories to ensure that no group has undue influence within the code and standards development process.

| **NFPA Member Categories** |
| :-: |
| Manufacturer |
| User |
| Installer/Maintainer |
| Labor |
| Enforcing Authority |
| Insurance |
| Special Expert |
| Consumer |
| Applied Research/Testing Laboratory |

**NOTE** NFPA committees are structured so that not more than one-third of any committee represents a single interest group.

The fact that a code provision has the approval of the majority of the membership, through a development process with rules that guard against undue influence by any one special interest group, sounds like a recipe for success. Unfortunately, some groups do have significantly more influence in the code process, although not through unethical or immoral means. Influence in the process is measured directly by participation. Participation requires money and time. Membership costs, travel expenses to attend hearings, and time away from the job all require using the limited resources that fire chiefs have to provide fire protection for the jurisdiction.

For the most part, the fire service has not been well represented in the code change process. Firefighters who have participated have performed admirably with few resources and against great odds. Industry recognizes the importance of participation on technical committees and considers the time and resources as the cost of doing business. But it is difficult for most fire chiefs to allocate resources to a national process involving cross-country air travel, rental cars, and hotels when budgets are strained and demands for service are increasing.

## Fire Service Commitment

With the voting procedures adopted by the ICC comes the opportunity for full fire service participation in the entire code development process. With that opportunity comes the obligation, not only to the public but also to present and future firefighters, who will ply their craft within buildings and structures designed and built to the codes. The fire service must exercise influence through groups such as the International Association of Fire Chiefs, the International Association of Fire Fighters (IAFF), and regional organizations. The IAFF must recognize that code development directly impacts firefighter safety.

Buildings are built to a minimum code. A prominent point in the public discussion of the major fire incidents of the past century has been an indictment of the building regulations or lack thereof that often played a role in the catastrophe, yet design professionals have rightly asked, "Where were you guys when the code was being developed?"

**NOTE** With that opportunity comes the obligation, not only to the public but also to present and future firefighters, who will ply their craft within buildings and structures designed and built to the codes.

> As a young firefighter, I studied Francis Brannigan's Building Construction for the Fire Service. In the preface to the first edition, he gave a charge to the potential fire officer: "For better or worse, these are the buildings that have been built, know them, do the best you can, safeguard your men, but be aware that many of these buildings were built to burn or collapse."[11]

Brannigan's text has been required reading and has had significant influence on the modern fire service. What was left unsaid by most fire service texts was that we did not have to simply accept the fact that buildings were not adequately designed and constructed to resist the effects of fire. In the past, we had the excuse that we were outnumbered and outgunned. That excuse is gone. We can do something other than complain. The fire service must dedicate itself to becoming a major player in the model code development process.

**NOTE** The fire service must dedicate itself to becoming a major player in the model code development process.

## CODE ADOPTION

The act of state or local governments of accepting a model code and giving it the effect of law is called *adoption*. Two basic methods are used based on the laws in effect within the jurisdiction. **Adoption by reference** is simply the passage of legislation that states that a specific edition of a certain code will be enforced within the jurisdiction. The document is mentioned by reference only, and copies of the codes must be purchased from the model code groups. In some localities, adoption by reference is not legally possible. Adoption by transcription is a legal requirement, and the code is republished, usually with a numbering system for each section or article that complies with the requirements of the jurisdiction.

> **adoption by reference** code adoption process in which the jurisdiction passes an ordinance that lists or references a specific edition of a model code.

Model code groups permit adoption by reference for free—they make their money by selling books. **Adoption by transcription** is a different matter. A license to republish can be negotiated, or the model code organizations can be contracted to publish a special edition that meets the requirements of the adopting jurisdiction, and they get to sell the codebooks.

**adoption by transcription** code adoption process in which the model code is republished as an ordinance by a jurisdiction.

## State and Local Adoption

Whether a code is adopted as a state minimum code that can be locally amended, adopted as a state **mini-maxi code** with no option of local amendment, or as simply a locally adopted code, there are legal requirements to ensure that adequate public notice is given and citizens and special interest groups are given an opportunity to be heard regarding the proposed codes. Mini-maxi codes are favored by most business interests and by developers because they create uniformity within the state, and all lobbying efforts can be directed at the state capital, not within each political subdivision. The fire service generally opposed mini-maxi codes because local control is reduced. One positive aspect of mini-maxi codes is that they promote training and political action on a statewide level, forcing fire service organizations to work together.

**mini-maxi code** code adopted at a state level that cannot be locally amended.

**NOTE** Mini-maxi codes are favored by most business interests and by developers because they create uniformity within the state.

## WHAT CODES CANNOT DO

The cycle of catastrophe followed by a public outcry of "there ought to be a law" is not new and will not end anytime soon. Sometimes a legitimate need for new regulations is recognized, and new code provisions are adopted. In 1985, a fire at the Valley Parade Soccer Stadium in Bradford, England, claimed 56 lives and injured 200. Investigators identified the large volume of combustible materials beneath the grandstands as a significant factor in the fire spread. The fire, caused by discarded smoking materials that ignited trash and rubbish beneath the bleacher seats, engulfed a 290-foot-long grandstand in less than 5 minutes and was televised live with the soccer match.[12]

**NOTE** The cycle of catastrophe followed by a public outcry of "there ought to be a law" is not new and will not end anytime soon.

The Bradford stadium fire identified a potential problem not clearly covered by fire code provisions. A provision regulating waste and combustible materials storage under grandstand seats was added to national model codes and exists today in both the *Uniform Fire Code* and *International Fire Code*.

In the aftermath of the 2003 nightclub fire at The Station in West Warwick, Rhode Island, many called for new codes to ensure that such a catastrophe could never occur again. The lack of sprinklers was rightly identified as a significant factor in the fire that claimed 100 lives, but existing code provisions in place in the state of Rhode Island could have prevented the incident if they had been followed. The *Rhode Island Fire Code*, based on NFPA 1, included provisions that addressed indoor pyrotechnic displays and foam plastic as an **interior finish**. All of the codes or laws in the world will not prevent tragedies from occurring if they are not obeyed, and the largest fire prevention bureau in the nation cannot inspect every building every day.

**interior finish**  the exposed interior surface of a building or structure, whether for acoustical, decorative, insulative, or fire protection purposes.

**NOTE** Existing code provisions in place in the state of Rhode Island could have prevented the incident had they been followed.

**NOTE** The largest fire prevention bureau in the nation cannot inspect every building every day.

In the wake of the Sofa Super Store fire that took the lives of nine firefighters in Charleston, South Carolina, the National Institute of Occupational Safety (NIOSH) issued more than 30 recommendations. None addressed a critical issue that was identified by Chief Mike Chiramonte (retired) as part of an investigative team assembled by the City. "If we have an example of any building anywhere that shows that fire prevention can directly save firefighters lives, then this one does."

## SUMMARY

Model code organizations were originally formed for the purpose of reducing property loss and protecting the lives of the American public. They have evolved over the years into organizations that promote the public safety and welfare through development of regulations, training, and advocacy. A major reorganization within the model code community occurred in 2003 when the three regional model code organizations voted to consolidate and form the International Code Council. The NFPA

expanded its mission to include the development of a building code, *NFPA 5000*™, and partnered with the IAPMO, ASHRAE, and WFCA to maintain the C3 Codes (although NFPA 5000 has not been adopted by any major city or state).

Fire service participation in the process has been limited because of the financial constraints of municipal budgets, but fire service groups such as the International Association of Fire Chiefs continue to work toward greater fire service involvement in the process. The ICC increased the potential influence of the fire service by increasing the number of potential voting delegates. To be enforced within a jurisdiction, model codes must be adopted on a state or local level.

For codes to be effective, there must be adequate education, enforcement, and public cooperation. The governing body must have the political will to support the enforcing agency.

## REVIEW QUESTIONS

PG 74 & 76  1. What is the name of the organization that represents the consolidated memberships of BOCA, ICBO, and SBCCI?

PG 78  2. Which industry was responsible for the first model construction code in the United States?

PG 74  3. What is the difference between a code and a standard?     LAW     GUIDELINE

PG 78  4. What was the original mission of NFPA in 1896?

5. Why are mini-maxi codes favored by business groups and developers?

6. Are there any positive aspects for the fire service with the adoption of mini-maxi codes?

## DISCUSSION QUESTIONS

1. Your request to attend code development hearings has been denied based on the city manager's belief that the potential benefit does not justify the expense. Aside from the ability to vote on codes changes, what are the benefits of attending national code hearings?

2. Model codes are developed as minimum requirements to provide an acceptable level of public safety and property protection. What methods can be used by local governments to justify the need for local code requirements that exceed the model building and fire codes?

## CHAPTER PROJECT

Identify the following regarding the codes that are in effect within your jurisdiction:

- Model building and fire codes
- Edition (year)
- Name of the agency responsible for development and promulgation
- Location (within the government) of the agency or board
- Legal enablement (law, regulation, or executive order) authorizing code adoption

## ADDITIONAL RESOURCES

In-depth information on many of the subjects discussed in this chapter can be found in the following texts and publications and at these Web sites.

Percy Bugbee, *Men Against Fire, the Story of the National Fire Protection Association* (National Fire Protection Association, 1971).

International Code Council at www.iccsafe.org.

"Investigation Report: Fifty-six Die in English Stadium Fire," *Fire Journal*, May 1986 (available in report form from the National Fire Protection Association).

National Fire Protection Association at www.nfpa.org.

## NOTES

1. Harry Chase Brearley, *Fifty Years of Civilizing Force* (New York: Frederick A. Stokes, 1916), page 42.
2. Ibid., page 22.
3. Ibid., page 78.
4. *National Board of Fire Underwriters, Pioneers of Progress* (New York: National Board of Fire Underwriters, 1941), page 125.
5. Albert Harkness, "Building Codes: A Historical Perspective," *Building Officials and Code Administrators Magazine*, 26 (March/April 1996), page 15.
6. Ibid., page 19.
7. International Code Council, Press Release, March 4, 2003, 5203 Leesburg Pike, Suite 600, Falls Church, VA 22041, Gretchen P. Hesbacher.
8. National Fire Protection Association, *NFPA 100 Years, a Fire Protection Overview* (Quincy, MA: National Fire Protection Association, 1996), page 9.
9. Leon Stein, *The Triangle Fire* (New York: Lippincott, 1962), page 24.
10. Victoria Neufeldt et al., *Webster's New World Dictionary* (New York: Pocket Books, 1995), page 130.
11. Francis L. Brannigan, *Building Construction for the Fire Service* (Boston: National Fire Protection Association, 1971), preface.
12. Tom Klem, "Investigation Report: Fifty-six Die in English Stadium Fire," *Fire Journal*, May 1986, page 126.

## CHAPTER 5

# Plan Review

## LEARNING OBJECTIVES

Upon completion of this chapter, you should be able to:

- Discuss the reasons construction plans are reviewed before construction.
- List the types of plans that fire departments typically review.
- List five potential site plan items for review.
- Discuss the benefits of fire department involvement in plan review for the business community.

# THE IMPORTANCE OF PLAN REVIEW

Within the grand scheme of ensuring public safety, construction regulation has traditionally been the responsibility of a government official generally referred to as the *building commissioner, building superintendent,* or just plain *building official.* The need for construction regulation is rooted in catastrophic collapses and conflagrations that occurred throughout history. Similar to most forms of government regulation, it takes significant public sentiment before elected officials act.

Although building officials traditionally have been responsible for ensuring that structures are constructed and maintained within guidelines prescribed by code, the fire chief is forced to deal with emergency incidents involving those same structures. When the questions: "Why were so many people killed or injured?" or "Why did the fire spread so quickly?" or "Why wasn't the fire department able to stop the fire?" are asked in the aftermath of a tragic incident, rightly or wrongly, public officials sometimes point fingers. In the aftermath of the Triangle Shirtwaist fire in which 147 factory workers were killed in 1911, Fire Department City of New York (FDNY) Chief Edward F. Croker responded to the public outcry to place responsibility for the tragedy. "The matter [the building was constructed with inadequate means of egress] is entirely within the discretion of the Building Department Superintendent."[1]

**NOTE** In the aftermath of a tragic incident, rightly or wrongly, public officials sometimes point fingers.

The review of building plans and specifications before construction has been an integral part of the building permit application process. Plans for the Asch Building, which housed the Triangle Shirtwaist Company, were reviewed and approved in 1900—despite the fact that the building lacked a sufficient number of staircases. The employee who had reviewed the plan in 1900 had risen to the position of superintendent of buildings by the time the fire occurred in 1911.[2]

Fire departments today are being included in the plan review schemes of many jurisdictions. In some cases, their inclusion was the result of a tragedy. Often, it has been at the request of a building official who wants to ensure that complex issues that potentially impact fire department operations at emergencies are properly addressed by personnel with adequate technical knowledge. Whether the request is a proactive outreach or merely an attempt to deflect criticism in the future, the effect is the same. In some cases, review of fire protection systems such as sprinklers and standpipes rests solely with the fire department. Features including fire department access roadways, structural fire protection, means of egress, and specific hazards (including fuels, hazardous materials, and special hazards) are also reviewed.

**NOTE** Fire departments today are being included in the plan review schemes of many jurisdictions.

**FIGURE 5-1**
Plan review increases safety by ensuring that conditions are evaluated by a second set of eyes.

The primary purpose of a plan review process is to verify that the proposed structure complies with the code before work begins. Plan review serves other useful purposes as well. The system provides an additional level of safety by ensuring that conditions are evaluated by another set of eyes. The plan reviewer evaluates the structure on a drafting table (Figure 5-1). Inspectors in the field then compare the structure with the approved plans and the code. Building permits universally contain a caveat that might well be considered the eleventh commandment of building safety: "Approvals are subject to final field inspection." No matter how thorough and competent, reviewers at a drafting table cannot be expected to anticipate every condition or contingency. Sometimes conditions in the structure "as built" go well beyond what could be anticipated on a blueprint.

**NOTE** The primary purpose of a plan review process is to verify that the proposed structure complies with the code before work begins.

**NOTE** No matter how thorough and competent, reviewers at a drafting table cannot be expected to anticipate every condition or contingency.

The 1922 *National Building Code* required the submission of "all necessary plans of such proposed work" and "detailed structural drawings" with the permit application.[3] The building official was obligated to approve or reject the plan "within a reasonable time."[4] The 1940 *Uniform Building Code*, developed and published by the Pacific Coast Building Officials Conference (later known as the International Conference of Building Officials [ICBO]) contained much more detailed requirements, requiring plans to include "plots of the property, strain sheets, stress diagrams computations and other necessary data." Plans had to be prepared on substantial paper or cloth and be drawn to scale.[5] Clearly, the code writers were well aware that building plans and specifications would be the objects of intense scrutiny after any catastrophic event. Plans on "substantial paper or cloth" could be expected to remain legible and thus prove to be extremely useful evidence within the courts.

## PLAN ISSUES OF INTEREST TO THE FIRE SERVICE

Fire department access roadways to structures; the locations of **fire department connections**, hydrants, and **fire command centers**; and access to building services as well as control features for fire protection systems all have a significant potential operational impact in the event of an emergency (Figure 5-2). Imagine the entrance to a hospital complex with a security gate that is too narrow for fire apparatus to enter. Consider

**FIGURE 5-2**
Fire department access is a key element evaluated during plan review.

the impact of a fire department connection 6 feet above grade that a beleaguered pump operator must reach using a ground ladder, all the while hearing the call for water because the building fire pump is not supplying the **standpipes**. The fact that fire apparatus is bigger than it was 20 years ago or that having to use a ladder to reach a fire department connection really is a big deal might not be readily apparent to a plan reviewer outside the fire department.

**fire department connections** external hose connections to supply fire protection water for automatic sprinklers and standpipes.

**fire command centers** fire resistance–rated rooms normally found within high-rise buildings and other large structures that contain controls for fire protection systems, building systems and utilities, and communications systems.

**standpipe** a system of piping, valves hose outlets, and allied equipment installed in a building to distribute fire protection water.

## Site Plan Issues

The first plan normally submitted for review is a **site plan**. Site plan approval is the first step in the regulatory process and may even be part of a rezoning. Approval by the jurisdiction constitutes authorization for construction of a structure or structures for specific uses and of a specific construction classification. The location of the structure on the property, building height and area, and road configuration (including parking) is determined at this stage in the process. If there is a time in the process that is conducive to requests from the fire department that might exceed code required minimums, it is at the site plan review. Problems brought to the developer's attention after the site plan has been finalized usually result in costs that exceed budget projections and are cause for conflict.

**site plan** a plan for proposed development; typically the first plan submitted to the jurisdiction that includes location of structures, occupancy and construction type, proposed roadway and parking facilities, and available fire flow.

**NOTE** Problems brought to the developer's attention after the site plan has been finalized usually result in costs that exceed budget projections and are cause for conflict.

At times, conditions are attached to the approval by the jurisdiction. Conditions may include buffers between proposed structures and existing development, green space, parking restrictions, and other requirements. Typical site plan items are shown in Figure 5-3 on the plan review checklist developed by Sonoma County, California. Site plan items typically include building construction type and location, fire department vehicle access issues, fire protection systems, and fire flow requirements.

↳ GALLONS/MIN

## COUNTY OF SONOMA
# DEPARTMENT OF EMERGENCY SERVICES

### FIRE SERVICES • EMERGENCY MANAGEMENT • HAZARDOUS MATERIALS

VERNON A. LOSH II, DIRECTOR

## CERTIFICATE OF OCCUPANCY CHECKLIST
## COMMERCIAL BUILDING

PROJECT: _____     DATE:_____

ADDRESS:_____     CONTACT _____

PERMIT NUMBER: _____     PHONE:_____

SCOPE OF WORK: _____

( ) **CONTACT LOCAL FIRE DEPARTMENT** – Offer to have their representative attend the final inspection.

( ) **DES CONSTRUCTION FILE** – Review conditions, letters, Verify compliance in field. Verify that fees are input and up to date.

( ) **REVIEW APPROVED PLANS** - Verify DES approval letter, plan requirements and general note conditions are met.

( ) **INSPECTION APPROVAL SIGN-OFFS** (at Site) –

    ( ) **Underground Hydro/Flush**

    ( ) **Aboveground hydro, Final**

    ( ) **Fire Alarm**

    ( ) **Fire Pump Acceptance**

    ( ) **Sprinkler Monitoring**

    ( ) **Special System** (ie…High Piled Stock, Dust Collection, LPG tank, Aboveground Storage Tank)

( ) **NOTIFICATION** – *Notify fire department dispatch and building occupants prior to testing*

( ) **8 ½x 11" SITE AND FLOOR PLAN** Detailed site and floor plan provided for Fire Department fire planning..

( ) **BUILDING ADDRESSING** *CFC 901.4.4.* Illuminated, building-12 inches, two sides, contrasting background, plainly visible and legible from the road. Located at highest elevation of building. 6 inch suite numbers on front and back doors.

( ) **SITE ADDRESSING** *SCC 13-48.* If 30 feet from street and/or multi-building complex, address posted at driveway entrances. Visible from both directions.

( ) **GATES** *SCC 13-38.* Manual = Knox key lock provided, Powered = Knox access override system provided. Gate opens inward, entrance 2 feet wider than lane serving gate. Located 30 feet from roadway to allow a vehicle to stop without obstructing traffic.

Page 1 of 5

2300 County Center Drive, Suite 221A, Santa Rosa, CA 95403 • phone (707) 565-1152 • fax (707) 565-1172

## FIGURE 5-3
Typical site plan items are included in Sonoma County's checklist. (Courtesy of the Sonoma County Fire Department.)

---

**CERTIFICATE OF OCCUPANCY CHECKLIST**
**COMMERCIAL BUILDING**

( ) **FIRE DEPARTMENT ACCESS** *SCC13-29.* Access road provided to within 150 of exterior portions of building, as measured by an approved route around the exterior of the building.

    ( ) **SURFACE** *SCC 13-30.* All-weather surface, .8 class II - grades >5%. 5-10% plus sealed surface. 10-15% plus 2" asphalt concrete.

    ( ) **WIDTH** *SCC 13-34.* Minimum 18 feet in width, with not less than 20 feet of right-of-way.

    ( ) **TURNING RADIUS** *SCC 13-32.* Minimum 40' inside turning radius.

    ( ) **VERTICAL CLEARANCE** *SCC 13-33.* Unobstructed vertical clearance of 15 feet.

    ( ) **GRADE** *SCC 13-31.* Maximum grade of 15%.

    ( ) **TURNAROUNDS** *SCC 13-36.* Approved turnaround provided where drive exceeds 150 feet in length.

    ( ) **FIRE LANES** *CFC 902.2.4.3.* Fire apparatus access roads noted as fire lanes in accordance with CVC 22658.

    ( ) **OBSTRUCTION** *SCC 902.2.4.4.* Access roads not obstructed ie... parking, dumpsters, building materials.

    ( ) **BRIDGES** *SCC 13-33.* Load = HS 20. Structure capability, weight/vertical clearance signs. Vertical clearance –15'.

    ( ) **TURNOUTS** *SCC 13-35.* One-way drives in excess of 500 feet. Turnout provided at midpoint of road.

    ( ) **SECONDARY ACCESS** – Provided when required.

( ) **FUEL MODIFICATION** *SCC 13-54.* Fuel modification plan required as specified in SCC section 13-54 for buildings constructed in specified vegetation areas.

( ) **WATER SUPPLY** *SCC 13-51.* Private fire protection water system storage installed per approved plan.

( ) **FIRE HYDRANTS** *CFC 903.42 & Appendix IIIB.* Fire hydrants are provided per approved plan.

    ( ) **Location** per approved plan. 18 inches from curb or bollards provided. Outlets facing street.

    ( ) **Hydrant type** is per approved plan. *CFC 903.4.2.*

    ( ) **Reflectorized blue marker** (s) provided. *CFC 901.4.3.*

    ( ) **Access** to hydrants is clear (ie...landscaping, equipment.) *CFC 1001.7.1.*

( ) **FDC's / PIV's / Detector Checks**

    ( ) **SIGNS** – permanent metal sign wired indicating the areas controlled by the valve. Check plans for areas controlled.

    ( ) **ACCESS TO VALVES/PIV/FDC.** *CFC 1001.7.1.* 18 inches from curb or protected by bollards, No obstructions.

    ( ) **VALVES OPEN AND SECURED** and/or monitored as required *CFC 1001.6.3 and 1003.3.1.*

( ) **TRASH ENCLOSURES** *CFC 1103.2.2.* Trash enclosures greater than 1.5 cubic yards not stored within five feet of combustible walls, openings or roof eave lines. Approved if sprinklered.

( ) **KNOX BOX** *CFC 902.4.* Provided in buildings with hazardous materials, alarm systems, sprinkler systems or hazardous processes.

**FIGURE 5-3**
continued

# Building Plan Issues

In addition to site plan review, many fire departments review plans for **structural fire protection** and separation or **fire protection ratings**, emergency egress, and fire protection systems. Building departments typically review all aspects of the building, including structural ratings, **construction classification**, and means of egress, even though many of the issues are also reviewed by the fire department. Fire service review generally provides a second set of eyes to compare the proposed structure with the code and to attempt to identify potential problems or conflicts ahead of time. Issues involving hazardous materials or processes often require a level of expertise that goes beyond architecture or civil engineering.

**structural fire protection** protection afforded to structural elements to resist the effects of fire; protection is generally provided through encasement with concrete, gypsum, or other approved materials.

**fire protection ratings** protection provided for building elements from the effects of fire expressed in terms of time.

**construction classification** classification of a building or structure into one of the five types or subtypes included in the model building codes based on types of materials and structural protection afforded.

Emergency evacuation and the establishment of areas of refuge, where those incapable of evacuation can be sheltered, must be evaluated with an eye toward fire department operations. Again, the plan reviewer with fire operations experience or the civilian plan reviewer with an in-depth knowledge of fire department operations is a valuable asset that can complement the review by the building official's staff.

**NOTE** The plan reviewer with fire operations experience or the civilian plan reviewer with an in-depth knowledge of fire department operations is a valuable asset that can complement the review by the building official's staff.

# Fire Protection System Plan Issues

Perhaps nowhere in the plan review process is fire department input more valuable than in the review and approval of sprinkler, fire alarm, smoke control, and other fire protection systems. Fire operations personnel become the end users of the systems. There are numerous instances within the model codes and referenced fire protection system standards in which fire department approval is specifically required. The location of fire department connections and the configuration and access to fire command centers are just two examples. It is preferable to get fire department approval within a formal framework, similar to the plan review process. Asking for opinions from numerous operations personnel may lead to inconsistent decisions regarding what works best operationally.

# RESISTANCE TO FIRE DEPARTMENT PLAN REVIEW

Who could possibly be opposed to fire department involvement in plan review? Possible people who may be opposed may include design professionals, developers, contractors, elected officials, the building department, and perhaps even the fire chief! Each group has different concerns, and many are valid issues that must be addressed before the fire department ever becomes involved in the process. Fears that the plan review process will take longer and become more convoluted if the fire department gets involved are legitimate. Another concern of business interests is that fire officials will ask for anything they believe will improve safety, disregarding the code, and will attempt to blackmail developers into spending extra money to achieve an enhanced level of safety. Building officials may think the fire department is unnecessarily intruding into their territory. Fire chiefs may view the prospect as a method to siphon away already thin resources, coupled with another political headache. Each of these concerns is legitimate, and unfortunately, in some circumstances, rooted in fact.

## Resistance from Design Professionals and Developers

Time is money, and none of us has enough of either. Informing the business community that it will now take an additional 2 weeks to process a building permit will result in complaints to elected officials. When design professionals and developers find out that the additional time is needed to permit the fire department to review the plan, the complaints may increase. Design professionals may feel that they must jump through an additional hoop. Didn't the building official already check plans for fire protection? Nonetheless, it is hard for the business community to complain when they are informed that fire department involvement does not mean additional requirements. Rather, it means another set of eyes on the same issues and perhaps more consistency relating to fire department approvals.

Care must be taken to do whatever is necessary to streamline the plan review process. The model codes specifically state that they are minimum acceptable levels of protection, but fire officials must use the bully pulpit with care. Fire officials should strive for the safest buildings and promote increased levels of protection but do not put architects and engineers on the defensive. The development of formal plan review procedures and guidelines that are readily available for download from the department's Web site (Figure 5-4) will help to allay fears that the fire department will make unrealistic demands.

**NOTE** Care must be taken to do whatever is necessary to streamline the plan review process.

# Code
## Reference
### Package

**2006 Code Edition
May 2008**

## for
### Architects
### Engineers
### Designers
### Installers

**Fire Prevention Division
Engineering Plans Review Branch
10700 Page Avenue
Fairfax, Virginia  22030
Tel:  703 246-4806
Fax:  703-691-1053
TTY:  703-385-4419
http://www.fairfaxcounty.gov/fr/prevention/code_ref_pkg_06.pdf**

**May 2008**

**FIGURE 5-4**
Publishing concise plan review guidelines allays developer fears of "additional" or unnecessary requirements.
(Courtesy of the Fairfax County Fire and Rescue Department.)

## PLAN SUBMITTAL INFORMATION MATRIX

| Plan Type | Primary Code Reference | Submit Plans To | Phone Contact | Is a Permit needed/type? |
|---|---|---|---|---|
| Assembly/Exhibit. | Fire Prevention Code | FMO/Inspections | 703-246-4849 | FPCP@FMO |
| Building | USBC | DPWES/BPR | 703-222-0114 | @DPWES |
| Building Tenant | USBC | DPWES/BPR | 703-222-0114 | @DPWES |
| Fire Alarm | IBC 907 | FMO Plans Review | 703-246-4806 | Low Volt/F.AL/FMO |
| Fire Alarm Tenant | IBC 907 | FMO Plans Review | 703-246-4806 | Low Volt/F.AL/FMO |
| Fire Pump | NFPA 20-07 | FMO Plans Review | 703-246-4806 | Pump/FMO |
| Fireworks | Fire Prevention Code | FMO/Inspections | 703-246-4849 | FPCP@FMO |
| Foam | NFPA 11 Series | FMO/Plans Review | 703-246-4806 | Foam/FMO |
| Clean Agent | NFPA 2001 | FMO/Plans Review | 703-246-4806 | FMO |
| Special Locks | USBC 1008 | FMO/Plans Review | 703-246-4806 | Low Volt/F.AL/FMO |
| Propane (LPG) Tank | FPC/NFPA 58 | FMO/Plans Review | 703-246-4806 | FMO |
| Range Hood | IMC 509 | FMO/Plans Review | 703-246-4806 | FMO |
| Site Plan | PFM | FMO/Plans Review | 703-246-4806 | DPWES/OSD S |
| Sprinkler | 13-07 | FMO/Plans Review | 703-246-4806 | SPK@FMO |
| Sprinkler Tenant | 13-07 | FMO/Plans Review | 703-246-4806 | SPK@FMO |
| Tank Removal | FPC | FMO/Inspections | 703-246-4849 | FPCP@FMO |
| Tent/Temporary | IBC 3103 | FMO/Plans Review | 703-246-4806 | FPCP@FMO |
| Aboveground Tank | FPC/IMC | FMO/Plans Review | 703-246-4806 | @FMO |
| Underground Tank | FPC/IMC | FMO/Plans Review | 703-246-4806 | @FMO |

**FEES: All fees are calculated per the fee schedule in Chapter 61, Code of the County of Fairfax.** This includes work done in Plans Review, Systems Testing, and Inspections. Billing rate is per published fee schedule.

DPWES = Department of Public Works and Environmental Services
FMO = Fire Marshal's Office (Fire Prevention Division)
FPCP = Fire Prevention Code Permit

**FIGURE 5-4**
continued

## Resistance from the Building Official

The building official should see the fire department as an ally, not an adversary. It should be clear that the fire department is not usurping the building official's authority. Needless delays can result from plans being transported between agencies. Delays in permit processing may lead to political pressure. Establishing a fire department plan review office at the physical location of the building department saves considerable time and leads to improved communication among agencies. Disagreements should be settled behind closed doors. If developers see a rift, they may try to use it to their advantage.

**NOTE** The building official should see the fire department as an ally, not an adversary.

## Skills, Knowledge, and Ability for Plan Reviewers

Among the jobs that firefighters are called on to perform, a job in the fire prevention bureau requires skills and demeanor that do not necessarily match those we look for in firefighters. For some very good, very competent firefighters, a sometimes tedious job in a business setting might not be the best match. The job of reviewing construction plans and meeting with developers and design professionals is even further removed from the world of the fireground and fire station.

**NOTE** A job in the fire prevention bureau requires skills and demeanor that do not necessarily match those we look for in firefighters.

Design professionals and other code officials will rightly wonder what qualifies a member of the fire department to operate in their world. Some departments have the luxury of hiring engineers to serve as plan review staff. These civilian personnel should receive in-depth training on fire department operational objectives and operational needs. They must have access to senior officers within the department when operational expertise is needed.

**NOTE** Design professionals and other code officials will rightly wonder what qualifies a member of the fire department to operate in their world.

Firefighters also serve admirably as plan review staff in many jurisdictions. Training courses from the National Fire Academy; International Code Council (ICC); National Fire Protection Association (NFPA); and classes at technical schools, colleges, and universities are available.

## Certification

*NFPA 1031: Standard for Professional Qualifications for Fire Inspector and Plan Examiner* contains job descriptions for two levels of plans examiner, Plan Reviewer I and Plan Reviewer II. Both review building and fire protection system plans for compliance with applicable codes; represent the fire department at meetings with architects, engineers, and developers; and make recommendations for variances and code modifications. The Plan Reviewer II has additional experience and is tasked with training personnel at the Plan Reviewer I level.

The standard specifies what are considered "minimum standards" for professional competence in terms of job performance requirements (JPRs). The standard does not address management responsibilities, and as with all of the NFPA's standards for professional certification, it states that it is not the intent of the standard to restrict any jurisdiction from exceeding or combining the minimum requirements. The standards can be used to establish minimum requirements for those seeking assignment to plan review and to establish a list of knowledge, skills, and abilities (KSAs) to be used in selecting applicants for appointment. Many organizations require those appointed to obtain certification under the standard within a specified time frame.

National certification from the model code organizations is also available and shows that the fire department official is serious about performing well. The consolidation of the Building Officials and Code Administrators, International, International Conference of Building Officials, and Southern Building Code Congress International into the ICC has led to a merger of each group's certification registers. Testing and recertification are now accomplished through the ICC. Fire department plan review personnel can obtain the same certifications as their peers in the building department. Firefighters performing plan reviews should have to meet the same standards as members of the building official's staff. In addition to the JPRs contained in *NFPA 1031: Standard for Professional Qualifications for Fire Inspector and Plan Examiner*, other KSAs include:

- Ability to read blueprints
- Technical math skills
- Ability to communicate orally and in writing
- Ability to interpret complex technical codes
- Good mechanical aptitude
- Knowledge of basic construction principles
- Demonstrated firmness, fairness, and flexibility
- Impeccable honesty and forthrightness
- Willingness to serve as a member of a team

## National Institute for Certification in Engineering Technologies

The National Institute for Certification in Engineering Technologies (NICET) is a nonprofit division of the National Society of Professional Engineers. Founded in 1961, the NICET's mission statement reads:

> Provide an independent evaluation of technical knowledge and experience, through certification, among those working in the fields of engineering technology; define and support career paths for engineering technologists and related disciplines; and ensure recognition and continued professional development of certified individuals.

The NICET maintains technical certification programs for civil, mechanical, and electrical technicians in construction, geotechnical, transportation, utilities construction, security systems, communications, and fire protection. Fire protection technicians are certified through testing, verification of work in the field by immediate supervisors, employment history, and personal recommendation. The NICET certifies fire protection technicians in the following disciplines:

- Fire Sprinkler System Layout
- Fire Alarm Systems
- Inspection and Testing of Water-Based Systems
- Special Hazards Suppression Systems

Many state and local governments require NICET certification in order to design, repair, perform maintenance, or inspect fire protection systems. NICET certification is also the gold standard for employers who hire fire protection technicians. Code officials involved in plan review and inspection of fire protection systems should consider NICET certification. Seminars held by state and regional associations that prepare candidates for NICET examinations are a wise investment.

## SUMMARY

Fire department involvement in the plan approval process has significant potential that extends over years. Attention to detail during design results in benefits over the life of a building or structure. Fire department operations expertise in the review of fire department access roadways, structural fire protection, means of egress, and specific hazards (including fuels, hazardous materials, and special hazards) can resolve potential problems in the early stages of development. Failure to address problems or operational impediments before the beginning of construction operations often results in the fire department's living with significant problems for the life of the building.

## REVIEW QUESTIONS

1. List three types of plans that fire departments typically review.
2. List three potential site plan items of particular concern to fire departments.
3. List three objections to fire department involvement in the plan review process voiced by architects, developers, and other public officials.
4. List three benefits that counter those objections.

## DISCUSSION QUESTIONS

1. Develop a list of KSAs for inclusion in a job announcement for a new plans examiner position within your fire prevention bureau. Use *NFPA 1031: Standard for Professional Qualifications for Fire Inspector and Plan Examiner*, Appendixes D and E for guidance.

2. Use the Internet to find training classes for firefighters who are assigned to perform plan review. Develop a list of classes and associated tuition costs. How would you justify the expenditures to the fire chief and city council?

## CHAPTER PROJECT

- Make a flowchart of the permit and plan review process used by the regulatory agency or agencies of your jurisdiction.
- Identify redundancy or duplicative effort within the process.
- Develop a short argumentative statement supporting the elimination of redundancy and in favor of streamlining the process.
- Develop a short argumentative statement that justifies redundancy for increased public safety, supporting a thorough review that ultimately benefits the public and the building owner.

## ADDITIONAL RESOURCES

In-depth information on many of the subjects discussed in this chapter can be found in the following texts and publications and at these Web sites.

*Building Department Administration*, Robert O'Bannon et al. (International Code Council, 2007).

Howard Dean, *Legal Aspects of Code Administration* (International Code Council, 2002).

The Fairfax County Fire and Rescue Department's *Code Reference Manual for Engineers, Architects and Designers* is available at www.fairfaxcounty.gov.

Information on additional training for plan review personnel is available at the International Code Council's Web site at www.iccsafe.org.

*NFPA 1031: Standard for Professional Qualifications for Fire Inspector and Plan Examiner* is available at the National Fire Protection Association's Web site at www.nfpa.org.

## NOTES

1. Leon Stein, *The Triangle Fire* (New York: JB Lippincott, 1962), page 118.
2. Ibid.
3. *National Building Code*, 4th ed. (New York: National Board of Fire Underwriters, 1922), page 8.
4. Ibid., page 10.
5. *1940 Uniform Building Code* (Los Angeles: Pacific Coast Building Officials' Conference, 1940), page 21.

## CHAPTER 6

# Inspection

## LEARNING OBJECTIVES

Upon completion of this chapter, you should be able to:

- Define the term *acceptable risk* and discuss its impact on the level of code enforcement.

- Discuss the potential impact of inspections on a jurisdiction's fire record.

- Define the term *selective enforcement* as applied to fire codes.

- Describe two methods of determining inspection priorities and discuss the advantages and disadvantages of both.

# THE VALUE OF INSPECTIONS

> *The heart of any fire hazard law is INSPECTION, the value of which is generally underestimated as it is not spectacular or colorful.* ["Report of the Virginia Advisory Legislative Council to the Governor and the General Assembly of Virginia," Senate Document No. 11, 1948]
>
> *The Commission recommends that local governments make fire prevention at least equal to suppression in the planning of fire department priorities.* [America Burning: Report of the National Commission on Fire Prevention and Control. *Washington, DC: U.S. Government Printing Office, 1972, page 167.*]

On February 20, 2003, fire erupted at The Station nightclub, in West Warwick, Rhode Island during a rock concert. Indoor pyrotechnic devices ignited foam plastic material that had been installed for acoustic purposes. The initial toll of 97 fatalities rose to 100 within 3 months, making it the fourth deadliest nightclub fire in U.S. history. State and local governments across the country scrambled to determine their vulnerability to a similar catastrophe. Fire chiefs and fire marshals were summoned to testify before elected officials, often in televised sessions. The elected officials wanted reassurances that a system was in place to prevent such a large loss of life in their jurisdictions. Many fire officials were forced to admit that a comprehensive inspection program that included evening and weekend checks of public assembly occupancies did not exist. Few, if any, fire officials were given the opportunity to mention that the reason inspections were not being conducted was because of the lack of staff resources—resources that can only be provided by the elected officials.

## Demand for Inspections

The level of support from the public and elected officials for the inspection process is a constantly evolving one. **Acceptable risk**, which is a measure of the level of the hazard potential that the public is willing to live with, constantly changes. Incidents like The Station fire result in a public demand for government officials to take some action that will provide assurance to the public. What elected officials want to hear is why the catastrophic event that has aroused the public concern could not have happened in their community. What they usually get is what will be done in the future to keep the same thing from happening in their community. The two are not the same. Government programs are very often reactive, not proactive. Government agencies have finite resources that must be allocated to achieve the greatest impact in support of their stated mission. Resources are often stretched thin.

**NOTE** The level of support from the public and elected officials for the inspection process is a constantly evolving one.

**acceptable risk**   the level of fire risk that the general public is willing to bear at a given time.

**NOTE** What elected officials want to hear is why the catastrophic event that has aroused the public concern could not have happened in their community.

Routine inspections by the fire service range from courtesy residential fire safety surveys by in-service companies to specialized inspections for code compliance in business and industrial occupancies. The preservation of property and protection of the public is the fundamental mission of the fire service. The reduction of hazards through the inspection process is one of the most effective means of accomplishing the mission.

The inspection process is truly underestimated and often maligned by some firefighters who would rather fight fires and by some in fire service management who cannot see past the allocation of resources for training and staff or the potential for political conflict caused by enforcement. Fire inspections reduce the rate and severity of hostile fires. To most people, the question, "Just how many fires does an effective inspection program prevent?" is strictly rhetorical and simply cannot be answered. However, the effectiveness of inspections programs has been statistically proven.

**NOTE** The effectiveness of inspections programs has been statistically proven.

In 1978, the results of a federally funded study were published under the title *Fire Code Inspections and Fire Prevention: What Methods Lead to Success?* by the National Fire Protection Association (NFPA). The study was cooperatively conducted by the NFPA and the Urban Institute, with funding and support from the U.S. Fire Administration, the National Science Foundation, and 12 metropolitan fire departments from across the United States. The study was undertaken to determine whether some fire code inspection practices actually resulted in fewer fires, lower fire loss, and fewer civilian casualties. The study focused on properties covered by fire codes, excluding one- and two-family dwellings.

The study used fire data from 17 cities and one metropolitan county and grouped fires into three categories based on the potential for intervention through the inspection process. The categories were (1) fires caused by visible hazards; (2) fires caused by foolish or careless actions; and (3) fires caused by incendiary, suspicious, or natural causes. The study determined that fires caused by visible hazards that can be directly remedied by inspectors were responsible for only 4 to 8 percent of all fires. Fires caused by carelessness, foolish actions, or electrical or mechanical failures (sometimes caused by lack of maintenance) accounted for 40 to 60 percent of all fires. The remaining fires (32 to 56 percent) were identified as incendiary, suspicious, or natural-cause fires and were considered unpreventable by inspection.[1] When the following information was

**FIGURE 6-1**

Fire rates were lower in jurisdictions that used fire suppression forces to conduct inspections.

compared with inspection practices, the findings underscored the effectiveness of routine fire inspections:

- Fire rates appeared to be significantly lower in cities that annually inspected all, or nearly all, public buildings (Figure 6-1). Jurisdictions that did not annually inspect most public buildings had rates of fires exceeding losses of $5,000 that were more than twice that of those that did inspect them.

- Cities that used fire-suppression companies for a large share of their routine fire code inspections had substantially lower rates of fire, presumably because cities that exclusively used fire prevention bureau inspectors did not have sufficient personnel within the bureau to make annual inspections at all public buildings.

- Fires caused by carelessness or foolish actions or electrical or mechanical failures (40 to 60 percent of all fires) were not preventable through direct action by inspectors because these causes are not readily visible during an inspection. Yet the rate at which these fires occurred was significantly lower in cities with regular inspection programs, indicating that departments that inspect more frequently have more opportunities to motivate occupants. This finding also underscores the value of company inspections, recognizing the fact that although company inspections cannot be as in depth because of the differences in training between inspectors in the bureau and those with other responsibilities, inspections by in-service companies have a significant impact on the incidence of fire within the jurisdiction.

In a nutshell, the study showed that although only 4 to 8 percent of fires were caused by conditions that inspectors could detect and take action to mitigate the hazard (Figure 6-2), the incidence of fire could be reduced by 50 percent by undertaking a comprehensive program in which every public building is inspected. Inspection is a positive motivating factor for the prevention of fire.

**FIGURE 6-2**

Only 4 to 8 percent of fires were caused by conditions that inspectors could detect and take action to mitigate the incidence of fire could be reduced by 50 percent by undertaking a comprehensive program in which every public building is inspected.

**NOTE** The incidence of fire could be reduced by 50 percent by undertaking a comprehensive program in which every public building is inspected.

**NOTE** Inspection is a positive motivating factor for the prevention of fire.

# THE INSPECTIONS PROGRAM

The case for inspecting every occupancy, with the exception of dwellings, is compelling. Unfortunately, in most jurisdictions, a finite number of inspections can be conducted by the fire prevention bureau. Many jurisdictions enlist the aid of in-service companies to increase the number of inspections. The 1979 NFPA study clearly demonstrated the effectiveness of company inspections even though most station personnel had not received the same level of training as fire prevention bureau personnel. If the bureau can only perform a percentage of the inspections, which occupancy classes warrant the added training and expertise? Inspection priorities must be established. Which occupancies should be inspected by the bureau? How often should inspections be conducted? How can priorities be established that will be fair, based on demonstrated need, and not lead to complaints of selective enforcement by certain businesses and groups?

## The Purpose of Fire Inspections

At times, the basic issues involving the establishment of a public program are obscured by the sheer volume of information that has to be considered. It is pretty clear that the desired result of any fire prevention program is fewer fires, reduced fire loss from fires that do occur, and fewer deaths and injuries. The 1975 American Insurance Association's *Special Interest Bulletin No. 5, The Value and Purpose of Fire Department Inspections*, listed the following seven objectives for an inspection program, which when accomplished, would render a substantial service to the community served.[2]

1. To obtain proper life safety conditions. Life safety inspections warrant attention to the adequacy of exits, obstruction to adequate and orderly egress at the time of fire, the adequacy of building evacuation plans, and determination of the number of persons permitted in places of assembly.

2. To keep fires from starting. Persons who work among hazardous materials or situations often become negligent of their own safety, just as long periods without a fire can cause overconfidence and underestimation of the fire problem.

3. To keep fires from spreading. The general public has little appreciation of the great value that structural features such as stair and elevator enclosures, fire doors, and fire partitions have in preventing the spread of fire.

4. To determine the adequacy and maintenance of fire protection systems.

5. To preplan firefighting procedure. It is difficult to attack a fire intelligently without first knowing the building and its occupancy (Figure 6-3).

**FIGURE 6-3**
Company inspections afford an excellent opportunity to preplan.

**FIGURE 6-4**
Establishing good relations with the business community promotes cooperation.

6. To stimulate cooperation between owners and occupants and the fire department. Nothing will ensure closer cooperation between owners and occupants of buildings and the fire department than the interest of the department in not only preventing fires but also in being better prepared to handle fires when they occur (Figure 6-4).

7. To ensure compliance with fire protection laws, ordinances, and regulations. In many places, some or all of these matters are under the jurisdiction of the fire department.

## Selective Code Enforcement

✳ ✳ **Selective code enforcement** is a term describing inspection and enforcement efforts ✳ ✳
that are not based on legitimate factors and that use the power of the government to
selectively discriminate against a particular business or group. As distasteful as it may
seem, fire officials must always be on guard against influences that might lead to the perception of selective enforcement. Attempts have been made to use fire and building codes
to keep group homes for mentally challenged individuals out of certain neighborhoods.
Calls for strict enforcement in certain religious establishments, in certain industries, or
in certain neighborhoods are sometimes thinly veiled attempts to use the fire inspections
process to intimidate or drive out a particular group.

| selective code enforcement | illegal code enforcement actions based on political or other motives. |
|---|---|

**NOTE** Fire officials must always be on guard against influences that might
lead to the perception of selective enforcement.

Selective code enforcement is <u>unethical</u> and <u>illegal</u>. Fire officials should not be
shocked during the course of an enforcement action when accusations of selective
enforcement arise. For some, the threat of such accusations is perceived as a silver bullet
that will force fire officials to back away from performing their duty. But the accusations
are sometimes, unfortunately, based on the bad experiences from the past. Inspections
programs should have written guidelines that plainly state inspections priorities and
goals. Enforcing the code is the right thing to do. Allowing the unsafe condition to occur
in violation of the code because of political pressure is a violation of the public trust and
is illegal.

**NOTE** Selective code enforcement is unethical and illegal.

## ESTABLISHING INSPECTION PRIORITIES

One of the initial steps in establishing an inspection program is determining which
occupancies will be inspected, how often, and in what order. In the aftermath of a tragic
fire in a particular type of facility, that group generally becomes the focus of public attention. Although understandable and not totally without merit, this system of "closing the
barn door after the horse has run off" does not lend itself to success. A well-disciplined
approach that prioritizes inspections based on risk and the actual hazards associated
with certain occupancy classes is by far the better course.

**NOTE** A well-disciplined approach that prioritizes inspections based on risk
and the actual hazards associated with certain occupancy classes is by far the better
course.

A system of inspections based on the actual fire experience within the jurisdiction would appear to be the surest method of addressing the most pressing fire prevention and life safety issues. An accurate assessment of fire statistics for the jurisdiction should be a regular function of the fire prevention bureau. Unfortunately, the results of the study of a statistical sample as small as most cities or counties does not reveal a true picture of the relative risks associated with particular hazard classes.

For the purpose of discussion, let us assume you have been tasked with determining inspection priorities, and in the course of your study of fire causes within your jurisdiction, you determine that welding and cutting are insignificant fire causes with negligible losses for the statistical period. Welding and cutting might go to the bottom of the list of inspections, if it made the list at all. In reality, it is a serious fire cause, with significant losses. The *International Fire Code Commentary* described losses by a single insurance company within a 5-year period of 290 hot-work–related fires, with an average $1.4 million loss per incident. Of the 290 losses, 42 percent were caused by employees, and 58 percent were caused by outside contractors. The same insurance company experienced 395 fires associated with housekeeping and 262 losses associated with smoking. The average poor housekeeping loss was $902,000; the average smoking loss was about $440,000.[3] Sooner or later, welding and cutting would rightly assume its place in your jurisdiction's fire statistics and would become an inspection priority— after the damage is done. What method can be used to develop priorities that reflect actual hazards? The method is actually an integral part of every model fire code, the **permit system**.

**permit system** system of determining inspection priorities based on permits required by the model fire codes.

## The Permit Model

The 2009 *International Fire Code* contains 46 operational **permit** categories and 12 construction permit categories (Figure 6-5). Operational permit categories include special amusement buildings, carnivals, compressed gases, covered mall buildings, cutting and welding, explosives storage and use, and operation of lumberyards and woodworking plants. Construction permits include installation or modification of fire alarm and detection systems, flammable and combustible liquid tank installation or modification, installation or modification of spray rooms or booths, and erection of tents and air-supported structures. Permit categories across the different model fire codes are quite similar.

**permit** approval from the appropriate code official that authorizes construction, operation of a regulated process, or the maintenance of a regulated occupancy type.

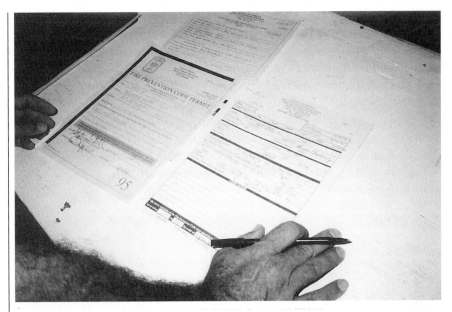

**FIGURE 6-5**
The permit system prioritizes inspections and protects against accusations of selective code enforcement. (Courtesy of Lionel Dukwitz.)

Most model fire codes "authorize" the fire official to require and issue permits. Older editions of the codes mandated their use as a method to ensure regulation of certain high-risk occupancies and processes. The 1981 *BOCA Basic Fire Prevention Code* F-900.2 contained permit requirements for dry-cleaning plants: "A person shall not engage in the business of dry cleaning without a permit obtained from the fire official, which shall prescribe the degree of hazard (high, moderate, low) of the system to be used."[4] The same code required the fire official to inspect the operation before permit issuance and authorized the revocation of a permit for failure to comply with the code or conditions of the permit.[5]

If you compare the list of permits from the model fire codes with large-loss fires from the past, you will see a direct relationship. Permit categories are based on our national fire experience and were established over the years in response to incidents involving large loss of life or significant financial loss from property damage. The list is constantly evolving. The requirement in each of the model fire codes to operate a "special amusement building," can be traced directly back to the 1984 Haunted Castle fire at Six Flags Great Adventure Park in Jackson Township, New Jersey.

**NOTE** If you compare the list of permits from the model fire codes with large-loss fires from the past, you will see a direct relationship.

**Special amusement buildings** are temporary or permanent structures for entertainment or amusement in which the means of egress are intentionally confounded or not apparent because of visual or audio distractions or are not readily available because of the nature of the attraction or mode of conveyance through the structure. Examples run from an attraction at Disney World on the high end to a carnival fun house in a tractor-trailer or even a Halloween maze put on by a local service club. In the Six Flags fire, eight of the approximately 30 visitors within the structure were killed when foam plastic material lining the interior of the structure was ignited. The combination of dense smoke, a rapidly spreading fire, and built-in confusion as to the location of exits within the structure was deadly.[6] In the aftermath, the memberships of the model code organizations voted to approve code changes regulating such structures and to ensure their inspection by requiring operational permits.

**special amusement buildings**   a permanent or temporary structure that provides a walkway or method of transportation wherein the required means of egress is not apparent or is confounded through the use of theatrical or special effects.

When faced with the prospect of inspecting a limited number of processes and occupancies, the decision to inspect those requiring permits under the fire code is a wise one. Not only are the facilities with the greatest risk for fire or life loss being afforded the attention of the fire prevention bureau first, the choice is based on the national fire experience. A particular industry's cry of "Why are you picking on us?" can be quickly answered by identifying the code requirement.

**NOTE**   When faced with the prospect of inspecting a limited number of processes and occupancies, the decision to inspect those requiring permits under the fire code is a wise one.

A significant but often overlooked benefit of the permit system is the leverage that permits offer in gaining compliance and gaining entry to inspect. Operation without the required permit is a violation of the model fire codes. Permits can be revoked for failure to comply with the code or violation of permit conditions. **Permit conditions** include submission of the required application, payment of fees, and permission for the fire official to inspect the facility.

**permit conditions**   code-specified conditions regarding permit issuance, including posting of the permit, submitting to inspection, accuracy in the permit application process, and complying with all code provisions.

The courts have consistently upheld the right of property owners, including business owners, to prohibit the entry of government officials without a search warrant. In other words, Bob the body shop owner has the constitutional right to refuse entry to a fire inspector. What Bob does not have is the constitutional right to a fire code permit—no inspection, no permit. Spray painting cars in his body shop without the required fire code permit for spray application of flammable finishes is a violation of the fire code.

Violating the fire code is a criminal misdemeanor in many jurisdictions. What might seem a bit heavy-handed is a useful method of gaining entry without the cumbersome process of securing inspection warrants. Rarely do fire officials have to use the muscle provided by the model fire codes to gain entry, but when faced with deliberate attempts to block the official performance of duty, the fire official should not hesitate to bring to bear the amount of muscle needed to get the job done.

## The Inspection Model

The other method of determining inspection priorities is the **inspection model**. Occupancy classes and types of processes are selected for inspection based on perceived need. The need must be based on fire risk or life hazard, and it should be backed up with statistics of actual fire loss. The fire official must be able to justify why particular industries, processes, or occupancies are selected and why others are not. Inspecting by district or area immediately raises the question, "Why were we selected first?" There may be a legitimate reason. The fire official must be ready to provide valid reasons to a skeptical media and public. The elected official who presses the fire official to "clean up" a particular part of the city for other than legitimate reasons will disavow any knowledge of the demand when the story makes the news.

**Inspection model** method of determining inspection priorities in which local officials select the occupancies to be inspected.

When prioritizing inspections using the inspection model, develop a written plan. If the need does arise to secure an **inspection warrant**, the judicial officer will want to know why the particular occupancy was selected for inspection to start with. A good answer would be: "Our standard operating procedure requires that all fuel storage facilities are inspected annually. We last inspected the facility 12 months ago," or "We inspect all fuel storage facilities every fall." The judicial officer represents the people and is supposed to make certain that your request for an inspection warrant is based on a legitimate government need.

**NOTE** When prioritizing inspections using the inspection model, develop a written plan.

**inspection warrant** administrative warrant; a warrant issued by a court with jurisdiction commanding an officer to inspect a specific premise.

## INSPECTIONS OF EXISTING OCCUPANCIES

Within the big picture of inspection and enforcement of codes by state and local governments, the fire official probably has the broadest mandate. Most jurisdictions have a building official charged with the enforcement of a construction code that regulates new construction and renovations. Many jurisdictions have a property maintenance code that targets residential occupancies, particularly rental properties, to ensure that land-lords meet minimum standards for safety and hygiene. Health inspectors are normally involved in food service and processing, health care, and other facilities that may have public health implications. In many states, assisted living facilities, foster care, and day care for both adults and children may be regulated by social service agencies. The fire official is generally the only official whose code applies to all of the foregoing and should interact with all the other officials.

**NOTE** Within the big picture of inspection and enforcement of codes by state and local governments, the fire official probably has the broadest mandate.

With very few exceptions, all buildings and structures, including residential occupancies, are within the scope of the model fire codes. What separates residences from other types of occupancies is that they are exempt from routine inspection. The **right of entry clauses** in the model codes that permit the fire official to enter and inspect structures on a routine basis exempt one- and two-family dwellings and residential portions of multifamily structures. The code still applies and must be enforced when conditions exist that pose a significant hazard. Just what type of hazard reaches the level of significant is subjective and has been debated among fire officials, civil libertarians, and the courts for years.

**NOTE** With very few exceptions, all buildings and structures including residential occupancies are within the scope of the model fire codes.

**right of entry clause** clause within the model fire codes that authorizes the fire official to enter buildings and premises for the purpose of enforcing the code with the permission of the occupant.

When Supreme Court Justice Potter Stewart wrote his famous opinion regarding obscenity in 1964, he stated: "I shall not today attempt to further define the kinds of material I understand to be embraced . . . but I know it when I see it. . . ."[7] Wrestling with just what constitutes a hazard that is significant enough to warrant government intrusion within someone's home is similarly perplexing. The hazard must be sufficient to justify what our forefathers wrote the Fourth Amendment to the Constitution to guard against. The condition of your 12-year-old's room probably does not reach the level of hazard, but 55 gallons of lacquer thinner stored in the ground floor apartment of a painting contractor certainly does. The fact that lacquer thinner is highly flammable and a leak in such a large container would pose a threat to all the residents of the building would probably lead the judicial officer to determine that the potential threat to the neighbors is greater than the possible harm caused by the intrusion into the painter's home. In most cases, inspection warrants can only be obtained after a request for entry to inspect has been denied by the tenant or property owner.

**NOTE** The hazard must be sufficient to justify what our forefathers wrote the Fourth Amendment to the Constitution to guard against.

# Routine Inspections

Most inspections of existing occupancies are considered routine inspections. They are conducted during normal business hours and may involve the issuance of a permit. Routine inspections include all occupancy classes. Whether an appointment should be made with the operator of the facility is a much-debated topic. Some advocate unannounced inspections to get an unvarnished look at the way the facility actually operates. A second benefit to unannounced inspections is that inspectors have greater flexibility in scheduling and can reduce travel time by clustering their work within particular areas.

Advocates of scheduled inspections argue that inspections are not meant to be surprise maneuvers to catch the guilty. If the property owner knows you are coming and uses the opportunity to get the premises in order, thereby reducing the inspector's need to perform follow-up inspections, are you not getting exactly what you are supposed to be after? An interesting wrinkle to the scheduled versus unannounced inspection debate is the growing popularity of self-inspection programs for certain occupancies. The programs are generally limited to lower risk commercial occupancies, such as business and light industrial, with good fire and code compliance records.

Checklist-type self-inspection forms such as that shown in Figure 6-6 are mailed or delivered with prepaid envelopes. Business owners are asked to complete and sign the forms and return them to the fire prevention bureau. Included with the checklist instructions is a statement that the program is voluntary, that a certain percentage (usually 5 to 10 percent) of the responses will receive a follow-up inspection by the bureau for quality control purposes, and that a lack of response will ensure an inspection by the fire prevention bureau. These programs take advantage of two basic principles that are often overlooked by those of us in the fire protection business who have become perhaps a bit jaded over time. First, most people want to do the right thing. They do not want to risk a fire and certainly do not want to lose their business. Second, given proper direction, most business owners will address and correct safety deficiencies that can easily be overlooked in the routine course of business.

**NOTE** Given proper direction, most business owners will address and correct safety deficiencies that can easily be overlooked in the routine course of business.

When inspectors identify code violations, they are required by the codes to issue a written notice of violation to the owner or operator of the facility. The model codes use the term **shall**, which has the legal connotation of must, meaning the inspector has no choice but to issue the written notice and follow up to ensure compliance. Most model codes require that notices include certain elements, such as the specific violation, code section, required corrective action, and date of the follow-up inspection. The fact that a **notice of violation** states that a follow-up inspection is scheduled 10 business days from the date of the original inspection does not mean the inspector has granted the business

The City of Concord has a self-inspection program for businesses which have a statistically lower risk of fire.

Your business has been selected to participate in this program.

Here's how it works:

Using this brochure as a guide, walk through your business.

Start outside the building and work in. Try to look as if you are seeing the business for the first time.

Look carefully for items which might pose a hazard.

Make note of those items and correct them as soon as possible.

After you have made any necessary corrections, return this form to:
**Concord Fire Department**
**Fire Prevention Division**
**24 Horseshoe Pond Lane**
**Concord, NH 03301**

Please feel free to make a copy of this form.

In addition to the items listed inside this brochure, there are several inspections which are required when applicable:

Fire extinguishers must be serviced every 12 months.

Sprinkler sysytems have an annual inspection requirement.

Installed fire alarm systems must be tested annually.

Installed hood and duct fire protection systems (the type typically found in restaurants) must be serviced every six months.

If you have any questions about any of these required inspections, or about any of the other code requirements, please call 225-8651 or visit our web site at: www.ci.concord.nh.us/fire

Concord Fire Department offers free training in fire extinguisher use and other emergency operations. Call or visit the web site for more information.

By participating in this program, you are helping to control operational costs for the City. Additionally, the Fire Department is better able to concentrate inspection efforts on businesses that pose a higher risk.

If your business is located within a residence please indicate. Y/N

If you have any questions about this program or would like assistance with the inspection, please call 225-8651.

Thank you for your time.

Brian J. Troxler
Fire Marshal

Name ____________________

DBA ____________________

Address ____________________

Telephone ____________________

Date ____________

Signature ____________________

**Concord Fire Department**

**Self Inspection Program**

Please complete this inspection form and return to the Fire Department within 30 days.

☐ Are compressed gas cylinders secured to keep them from falling?
If a cylinder falls and the top breaks off, it can become a missile!

☐ Are circuit breakers / fuses labeled in the electrical panel box?
If you need to turn the power off to a certain portion of the building or to a particular piece of equipment, the labeling will help. Firefighters need to know, too.

☐ Is the area around the furnace and water heater clear of combustible materials?
This area tends to be a collection place for all sorts of storage, especially file boxes or those seasonal things which are only used briefly. Often the storage accumulates during the summer when the furnace may not be operating. Care must be taken to keep this area clear.

☐ Do emergency exit lighting fixtures work properly?
These fixtures help light the way to exits during power failures. Push the test button to check for proper operation.

☐ Is a clear space of 30 inches maintained in front of all electrical panel boxes for access?
In an emergency, when you need to get to the electrical panel, you don't want anything in your way. If a fire starts in the panel it could spread to materials in front of it and then through the building.

☐ Is there clearance beneath sprinkler heads?
Storage must not impinge on sprinkler heads. A clear distance of 18" beneath the sprinkler head is required for proper operation.

☐ Are address numbers for your building clearly visible from the street, at least 4" high and of a contrasting color from their background?
Emergency responders need to be able to find your business in a hurry. Be sure the numbers can be seen from the street.

☐ Are exit signs illuminated?
If you have lighted exit signs, the light bulbs must be replaced when they burn out.

☐ Are there less than 10 gallons of flammable liquids in the building? If more, is the storage in an approved flammable liquids cabinet?

☐ Are exit doors unlocked during business hours ?
All exits must be operable without the use of a key whenever the building is occupied. This includes employees as well as the public.

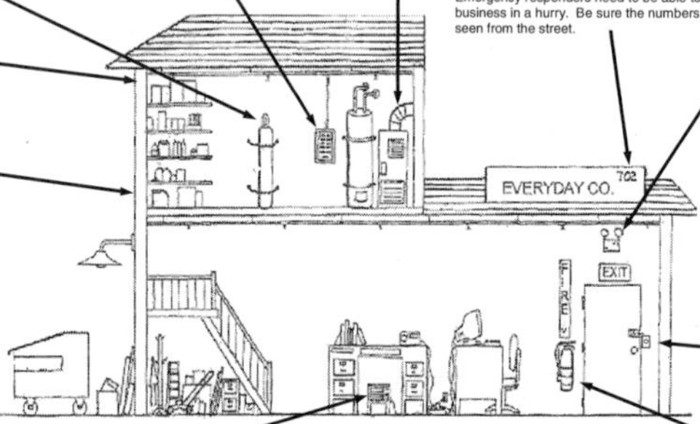


☐ Are dumpsters located more than five (5) feet from combustible wall, openings, roof overhangs, trees or fences on property lines?
If the dumpster catches fire and it is too close to the building, the fire will likely spread to the building. The same is true for all combustibles up against the building.

☐ Is the access to all exits clear?
Building occupants must be able to get to the exit doors without going around displays that are in the way. There should be no trip hazards. Nothing can be stored in front of an exit.

☐ Are portable electric heaters plugged directly into a wall outlet?
Extension cords may not be used (even if the have built-in circuit breakers). Heaters use a large amount of electricity and can melt an extension cord.

☐ Does the Knox Box have the proper keys?
If you have changed your locks, the keys need to be replaced in the Knox Box

☐ Are all fire extinguishers visible, Readily accessible and not blocked by storage or equipment?
Extinguishers need to be easily seen, clearly marked and mounted. If you can't get to an extinguisher easily, it won't be of any use.

☐ Do portable electric heaters have tip-over switches?
Portable heaters are frequently bumped into and knocked over. Heaters should have an internal safety switch that will turn the unit off if it tips over.

☐ Extension cords shall not be used in place of permanent wiring. If used on a temporary basis, are extension cords in good condition and do they have built-in circuit breakers?
Lightweight electrical extension cords are not allowed in businesses. If you must use an extension cord, it must have a built-in circuit breaker. These "plug strips" should not be used for large appliances or other equipment which draw a large load. These strips must always be plugged directly into an outlet and they cannot be "daisy-chained" together.

☐ Are portable heaters located a minimum of 36 inches away from combustibles?
Many fires are started when heaters are placed too close to boxes, drapes, furniture, etc. Keep a clear space around them!

☐ Has the fire extinguisher been serviced within the last 12 months?
There is an annual requirement for the maintenance of fire extinguishers. Check the tags on your extinguishers to see if they are current.

**FIGURE 6-6**
Self-inspection programs are used to supplement, not replace, fire department inspections. (Courtesy of the Concord Fire Department.)

owner a free pass to violate the code for 10 days. The inspector does not have that authority, and neither does the chief. The owner is obligated to remedy the code violation immediately.

> **shall** a positive and definitive requirement of the code that must be performed; action is mandatory.
>
> **notice of violation** written notice issued to a property owner or occupant listing (1) unsafe conditions, (2) applicable code sections, (3) required corrective action, and (4) date of a follow-up inspection to ensure compliance.

## Seasonal Inspections

Similar to most businesses, the fire protection business has busy seasons. The same seasonal conditions that create increased demand for fire suppression and subsequent fire investigation services increase the need for fire inspections (Figure 6-7). Certainly every fire cannot be prevented, but many can, or at the very least, the impact may be lessened through the effective inspection and code enforcement. The demand for fire suppression and investigation might have been reduced or eliminated if adequate resources had been brought to bear before the fire.

**FIGURE 6-7**
Inspections of holiday activities increase the workload of fire prevention bureaus, but they are predictable events.

Some holidays pose specific fire hazards that can be anticipated. Fourth of July fireworks, Halloween decorations in public assembly occupancies, and increased stocks of goods in mercantile occupancies during the Christmas shopping season all pose predictable hazards. Inspections programs must be flexible and address these hazards. The level at which the fire prevention bureau and the fire department is involved will very much be a function of potential risk and public demand.

**NOTE** Some holidays pose specific fire hazards that can be anticipated.

Holidays that promote revelry, including Halloween, Fat Tuesday, Cinco de Mayo, and even the Super Bowl, can sometimes lead to unsafe conditions because of increased occupant loads, highly combustible decorations, and alcohol consumption. Often the best method of preventing unsafe conditions from occurring is providing information beforehand to the business community outlining restrictions on combustible decorative material and a reminder that inspectors will enforce occupancy load limits. The worst time to try to make the point is at midnight in the middle of a crowded bar surrounded by people in Halloween costumes.

Fire prevention bureaus should develop formal organizational ties to other agencies such as law enforcement and other regulatory agencies, as well as business, religious, and other associations. In addition to sharing information of upcoming events, it is good customer service when a fire inspector is able to steer a citizen with a problem to a specific person in a specific agency for help in resolving a problem. It is even better customer service when the inspector makes the contact and has the other official contact the citizen.

## Fireworks

Fireworks, both consumer and aerial displays, are a major cause of hostile fires and injuries. Some jurisdictions have mercifully banned consumer fireworks, only to have their citizens order them from mail-order catalogs or on the Internet. In some jurisdictions, the public will not tolerate a ban, regardless of fires and injuries. In these jurisdictions, the best course is usually rigid regulation of fireworks sales and sure prosecution of violators.

**NOTE** The best course is usually rigid regulation of fireworks sales and sure prosecution of violators.

The model fire codes require permits for the manufacture, storage, sale, and use of fireworks. By using the permit process, fire officials can limit the location of sales areas, ensure proper fire protection, and make it unequivocally clear to operators that permits will be revoked if unsafe or noncompliant conditions are found. Regulations should be well established and clear to all involved.

Regulation of aerial fireworks displays in congested urban and suburban areas may require actual on-site supervision by a fire inspector, including a review of the site plan and proposed display (mortar sizes), an inspection of the firing area, fallout and spectator

**FIGURE 6-8**
Aerial fireworks displays must be fired by experienced personnel. Accidents are often tragic.
(Courtesy of Duane Perry.)

areas, display of operators' credentials, and inspection of the fallout area for unfired fireworks after the display. Fireworks sales, storage, and display all require fire code permits. The permit process is key to effective regulation.

Permit application for aerial displays should include site plans, inventories, operator qualifications, storage arrangement, security arrangements, and a rain date. Rigid conformance with the fire code and NFPA standards should be the minimum (Figure 6-8). There is a long litany of tragic accidents, often involving civic groups, including fire departments, in which inexperienced or intoxicated persons were killed or injured because of the lack of training and safeguards.

## Special Inspections

Some occupancies pose special hazards based on industrial processes, special circumstances of the occupants, or even the hours of operation. Occupancies that pose significant life or fire hazards, occupancies housing complex industrial processes, and those that operate at odd hours often require a different level of attention by the fire official. Extra training and specialization for inspectors is sometimes warranted. At times, a fire official may need to enlist the aid of outside experts. In any case, the fire official is still charged with enforcement of the code and cannot delegate that responsibility to some other entity without legal authorization.

## Nightclubs and Large Assemblies

Large assembly occupancies have significant potential for loss of life. The sheer number of occupants, people's unfamiliarity with the structure, the potential for large quantities of combustible materials, and the use of intoxicating beverages create an atmosphere where a relatively small fire can result in a catastrophic loss. Nightclubs, theaters, and other nighttime venues should receive a normal fire inspection sometime during the normal business day when few, if any, customers are present; lighting is sufficient to clearly see; and a representative can accompany the inspector and access all parts of the business.

**NOTE** Large assembly occupancies have significant potential for loss of life.

Additional inspections should be conducted during the evening hours when the facilities are operating and are at peak attendance levels. During evening nightclub inspections, life safety becomes the primary issue. Inspectors should check for fire department access to the structure; clear fire hydrants and fire department connections; overcrowding; and unsafe conditions, including unapproved open flame devices, unsafe displays or entertainment, and combustible decorative items.

**NOTE** Additional inspections should be conducted during the evening hours when the facilities are operating and are at peak attendance levels.

Theaters and athletic events generally tend to use the same seating and exiting arrangements, making things more predictable. Special effects and displays may necessitate additional scrutiny. Fairs; craft, gun, and boat shows; and similar events often operate in armories and convention centers. A review of each new layout and an inspection before the show's opening is often warranted. Building owners must be made to understand that it is their responsibility to police their tenants. Written policies regarding vehicle fuel indoors, black and smokeless powder, or other hazardous materials should be spelled out within the rental agreements for the facility. If 25 dealers each bring 1 lb of black powder and 20 lb of smokeless powder to a gun show, each has brought the maximum amount permitted for the building. Who determines who gets to stay? A little planning with the facility operator before the event will reduce the possibility that the fire marshal ends up looking like the bad guy for enforcing limits on vendors after the show is open.

## Institutional Occupancies

Facilities that house persons with a diminished capacity for self-preservation because of illness, medical or psychological impairment, or incarceration have code-mandated emergency planning requirements. Fire plans, fire drills, and employee training must be developed by the facility operators and reviewed and approved by the fire official. The term **institutional occupancy** is a broad one, with significant differences in the levels of code-required fire protection and life safety features. Fire officials must ensure that the fire safety strategy approved for a location is appropriate for the residents and the structure.

**institutional occupancy** an occupancy that houses persons whose capacity for self-preservation is limited or diminished by reason of age, infirmity, or confinement.

**NOTE** Fire officials must ensure that the fire safety strategy approved for a location is appropriate for the residents and the structure.

Building code requirements for nursing homes where residents are totally incapable of self-preservation are extensive and allow a **defend-in-place strategy** to be used in case of fire. The code recognizes that general evacuation is physically impossible. Assisted living facilities and board and care facilities, however, are designed for people who need supervision but are otherwise capable of appropriate, unassisted response to a fire. Assisted living facilities often resemble combustible, multifamily apartment buildings (Figure 6-9). Complete evacuation in the event of fire is the only prudent fire safety strategy. A fire official who approves a defend-in-place fire plan for an unprotected frame structure designed with complete evacuation in mind is approving a potential catastrophe—in advance.

**FIGURE 6-9**
Assisted living facilities often resemble apartment buildings and are designed for total evacuation in event of fire.

**defend-in-place strategy** design system used for structures housing persons in which the likelihood of evacuation is remote because of infirmity, disability, confinement, or age; it includes fire protection features such as fire resistance–rated construction and automatic sprinklers, intended to extend tenable conditions during extinguishment or rescue.

**NOTE** A fire official who approves a defend-in-place fire plan for an unprotected frame structure designed with complete evacuation in mind is approving a potential catastrophe—in advance.

## New Construction Inspections

In the big picture of codes, the job of ensuring that new structures are constructed in accordance with the model codes belongs to the building official. Renovations of existing structures must also comply with current codes and are treated much the same as new construction. Even in jurisdictions where the roles of fire and building officials are well defined, there are still areas of overlap that can lead to confusion or conflict without an established working relationship among agencies.

## Coordination to Ensure Fire Operations

One of the most important aspects of construction regulation concerns the fire department's ability to deal with fires and emergencies during construction. Both fire and building codes address the subject. At the very least, fire prevention bureaus should act as a conduit between the fire and building departments to ensure that building inspectors understand the needs of fire operations personnel regarding vehicle access, water supply, posting building addresses and street signs, and the storage of construction materials and construction waste.

**NOTE** One of the most important aspects of construction regulation concerns the fire department's ability to deal with fires and emergencies during construction.

Buildings undergoing renovation while partially occupied and occupied structures in close proximity to construction sites can result in people within the occupied structures (responsibility of the fire official) being impacted by the construction process (responsibility of the building official). Regulation for safety is never black and white. In some jurisdictions, fire inspectors enforce regulations regarding new construction, either under the provisions of the fire code or as an agent of the building official. In all cases, code enforcement agencies should speak with one voice. Differences should be worked out behind closed doors, not in public view.

## Fire Protection Systems Tests and Inspections

If there ever was an element of new construction regulation in which the fire service has a vested interest, it is the installation and acceptance tests for fire extinguishing, alarm, and other fire protection systems. Improperly designed or installed systems result in increased false or nuisance alarms, unnecessarily tying up fire department resources, and endangering firefighters and the public. Chapter 7, "Fire Protection Systems Testing," addresses acceptance tests for new systems and required maintenance and retesting of systems.

## SPECIAL HAZARDS

The modern fire codes address a staggering array of safety concerns. The thought that an inspector needs to be an "expert" in every field is ridiculous at best and even dangerous in the instance that an inspector thinks that effectively regulating a process is the same as safely conducting the process. Nonetheless, some processes do require special training for inspectors. In jurisdictions with the luxury of adequate levels of staffing, some specialization is possible, keeping in mind that no matter how specialized, every inspector must have a good working knowledge of the basics.

## Explosives and Blasting

If there is an area of code enforcement that requires a cool head, strong personality, good people skills, and a grasp of political realities, it is construction blasting and explosives regulation. Combine an unforgiving process, often involving days of noisy, dusty operations, with land development, in which the pristine forested areas or fields adjacent to occupied residential areas are cleared to make way for new buildings, and the patience and good nature of the surrounding landowners are sure to be tested.

**NOTE**   If there is an area of code enforcement that requires a cool head, strong personality, good people skills, and a grasp of political realities, it is construction blasting and explosives regulation.

Occupants of the existing homes often feel that theirs should have been the last home built, and they expected that the wooded area (that has been owned by a developer for years) would always be there. Now in addition to losing their view, noisy track drills, bulldozers, and dump trucks crawl over the site from dawn to dusk. Every afternoon, three long horn blasts are followed by a deep rumble, and then their house shakes! The crack in the basement floor (that has actually been there for 2 years) must have been caused by the blast last Tuesday. If citizens conclude that there is no effective regulatory control, public officials can rightly expect a public outcry (Figure 6-10). The better educated and more affluent the citizenry, the louder, better organized, and more prolonged will be the public response.

Procedures should be established for rapid response to citizen complaints—within hours, preferably the same day. Inspectors should be well versed in the code, blasting controls, and explosives regulations. Blasting contractors should be required to notify persons in the immediate vicinity well in advance and should be encouraged to conduct **preblast surveys** in an attempt to document and photograph existing structural blemishes and damage before blasting operations.

**FIGURE 6-10**
Construction blasting is loud, dusty, and often the cause for citizen complaints. (Courtesy of Howard Bailey.)

**NOTE** Procedures should be established for rapid response to citizen complaints—within hours, preferably the same day.

**preblast survey** inspection of the structures in the vicinity of future blasting operations; the surveys identify existing structural defects and damage and give advance notice to surrounding property owners.

A blaster certification program, in which blasters are tested and certified, goes a long way toward allaying public concerns. The sight of a uniformed fire inspector making regular visits to the site will also convey confidence that someone is watching. Regulations regarding the storage of explosives on construction sites must be rigidly enforced, particularly in this age of threats of terrorism. Accurate records of explosives deliveries, **shot records**, and **seismograph records** must be maintained and should be routinely checked by the fire official.

**shot records** records of blasting operations.

**seismograph records** records of ground vibration and airblast measured during blasting operations.

**FIGURE 6-11**
Fuel storage facilities range in size from several hundred to several million gallons.

## Fuel Storage Facilities

Fuel storage facilities range from a few hundred gallons in a temporary tank to millions of gallons in tank farms (Figure 6-11). Tanks are installed above ground, underground, and inside buildings and enclosures. In many jurisdictions, the regulation of underground storage tanks (USTs) is the responsibility of an environmental official. When the U.S. Congress mandated state regulation of USTs to reduce damage to the environment, many states were forced to pass laws identifying enforcement officials. Without local fire officials in every jurisdiction, some states had to identify someone else. In most cases, however, the local fire official performs routine inspections of existing facilities, even if tank installation is the responsibility of another agency.

Tank installation and maintenance are complex procedures. Tank installations involve a series of tests to ensure tightness and conformance with code and the manufacturer's installation instructions. Failure to conform to either can lead to leaks or tank failure. Training for fire inspectors involved in tank inspections is time and money well spent.

## COMPANY INSPECTIONS

Most jurisdictions do not have sufficient staff within the fire prevention bureau to inspect every occupancy, yet studies have shown that jurisdictions that inspect every occupancy realize reductions in fire loss. The use of in-service companies to perform inspections has been successful in many jurisdictions.

**FIGURE 6-12**
In-service company inspectors should perform a basic safety inspection.

Personnel should be trained to perform a basic safety inspection and directed to request assistance from the fire prevention bureau if they encounter imminent hazards (Figure 6-12). The assignment of fire prevention bureau inspectors to specific districts enhances communications between the bureau and the station inspectors and facilitates their use as a resource. Checklist forms are a useful method of ensuring consistent inspections, and they provide a record of the visit for the department and the business owner. Company personnel should be informed that the inspection is a combination preplan visit, safety inspection, and chance to interact with the community they serve.

**NOTE** Personnel should be trained to perform a basic safety inspection and directed to request assistance from the fire prevention bureau if they encounter imminent hazards.

The opportunity to preplan hazards, provide building familiarization, and the chance to establish cordial relations with the community can literally make in-service inspections the excuse that every chief wants to advertise the department. Company inspections training should be prefaced with the statement that the goals of the in-service inspection program are building familiarization to increase firefighter safety, elimination of fire hazards, and improved community relations.

> **NOTE** The goals of the in-service inspection program are building familiarization to increase firefighter safety, elimination of fire hazards, and improved community relations.

# RESIDENTIAL INSPECTIONS

Many jurisdictions in the United States have instituted registration and inspections programs for rented residential occupancies. For the most part, these programs have been driven by the desire of cities to prevent urban decay. As families leave cities for newer suburbs, the percentage of owner occupied dwellings typically decreases. In many older cities, large Victorian homes have been converted into apartments. In some cases, properties are converted without the benefit of a permit and with little regard to construction or fire code requirements. In many cases, absentee landlords interested only in short-term profits, with no intention of long-term ownership of properties, defer or delay maintenance. A vicious cycle emerges. Run-down properties lower the value and attractiveness of neighboring properties. Vacant and abandoned structures become havens for squatters and criminal activity. A slum is born.

Residential properties are within the scope of the model fire codes, although fire codes have limited effectiveness addressing the real problem. Fire codes contain provisions to deal with excessive waste accumulations, means of egress issues, electrical hazards, and problems with maintenance of fuel-fired appliances. But the model property maintenance codes do a better job. They address the same fire safety issues as the fire code but also include maintenance of the structure (e.g., siding, painting, masonry); plumbing, electrical, and mechanical conditions; and heating, light, ventilation, and interior sanitary conditions (e.g., mold and overcrowding).

In many cities, fire department inspectors are cross-trained and serve as property maintenance inspectors using the property maintenance code. In cities with separate property maintenance inspectors, fire inspectors typically work hand in hand with them. Why should the fire department be concerned about the maintenance of residential properties? Consider this: Of the reported structure fires in 2008, 403,000 were in residential occupancies and accounted for 78.2 percent of all structure fires. Of these, 291,000, or 56.5 percent, of all structural fires occurred in one- and two-family homes. Another 95,500, or 18.5 percent, of all structural fires occurred in apartments. Fires in residential occupancies resulted in 2,755 fire deaths, or 83 percent of all civilian fire deaths.[8]

## Residential Inspection Effectiveness

Gauging the effectiveness of any inspection program is time consuming and difficult. Chief Larry D. Lamb of the Niles, Michigan Fire Department wrote a series of papers on the residential inspections program established in his city as part of the Executive Fire Officer (EFO) program. The City of Niles' small size (5.5 square miles) and population (12,000 residents) allowed Chief Lamb to compare residential fires in owner-occupied dwellings with residential fires in tenant-occupied structures. His research identified a 46% decline in fires in tenant-occupied residential occupancies that were regularly inspected by the City from 1983 to 2003; the incidence of fire in owner occupied residential occupancies which were not inspected remained virtually unchanged.[9]

## INSPECTION TOOLS

For years, tools have been developed to assist inspectors and to ensure consistency in the process. Checklists are common, and with the advent of computerized inspection programs, handheld data terminals, and personal data assistants (PDAs), checklists are now available in digital format. There is a danger, however. Checklists, whether paper or electronic, are useful aids in the inspections process; they are not substitutes for training and experience. The notion that an untrained and unskilled inspector can perform an adequate inspection by using technology is wrong.

**NOTE** The notion that an untrained and unskilled inspector can perform an adequate inspection by using technology is wrong.

## Paper Checklists

Code-based checklists can be easily developed within most fire prevention bureaus. The process of developing checklist items provides a good training opportunity. Inspectors should continue to carry codebooks and should refer to the code before issuing a notice of violation. You cannot enforce a checklist!

## Inspection Software

Inspections programs are available from a number of software developers (Figure 6-13). Consider these points before purchasing any program:

- How well established is the developer, and what platform is used for the program? Your system will require updates and modifications each time a new code is adopted.
- Is the program based on your code system? Can changes and updates be performed in house by bureau members?
- Does the system interface with department programs?
- Will the technology meet the legal requirements of your code? Most codes require leaving a printed copy of the report and copies of violation notices.
- Is the system user friendly?
- Can you contact another agency that has used the program?

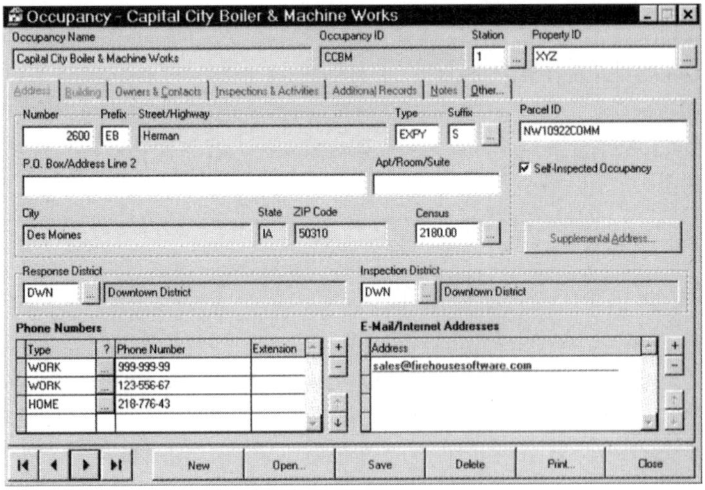

**FIGURE 6-13**

Commercial inspections software such as Fire House can be used by bureaus and in-service companies. (Courtesy of Fire House Software.)

If the answers are not forthcoming and clear, you might be better off with paper and clipboards. Chapter 11, "Fire Prevention Records and Record Keeping," contains a discussion of the legal requirements for fire prevention bureau records and records systems. The system must comply with your state regulations. Also consider how any program will interface with your jurisdiction's information management system. Many cities and counties have complex geographic information (GIS) systems that include property identification and information on taxes, public utilities, and infrastructure and enable inspectors in the field to access real-time property information. The most important aspect in evaluating any system is whether the system will be a time saver or just another complication that diverts staff efforts away from the real issue at hand. If a new system will force you to change the way you do business, and your way of doing business is successful, be wary of "stepping off a cliff."

## SUMMARY

The level of support from the public and elected officials for the inspection process is constantly evolving. "Acceptable risk" is the level of the hazard potential that the public is willing to live with without demanding government action. The preservation of property and protection of the public is the fundamental mission of the fire service.

The reduction of hazards through the inspection process is one of the most effective means of accomplishing the mission. The effectiveness of inspections programs was statistically proven in 1978 in a federally funded study published under the title *Fire Code Inspections and Fire Prevention: What Methods Lead to Success?*

One of the initial steps in establishing an inspection program is the determination of which occupancies will be inspected, how often, and in what order. When faced with the prospect of inspecting a limited number of processes and occupancies, the decision to inspect those requiring permits under the fire code is a wise one. Not only are the facilities with the greatest risk for fire or life loss being afforded the attention of the fire prevention bureau first; the choice is also based on the national fire experience.

## REVIEW QUESTIONS

1. In the 1978 study published under the title *Fire Code Inspections and Fire Prevention: What Methods Lead to Success?*, what percentage of fires was caused by conditions that could be observed during inspection?
2. What is meant by *selective code enforcement*?
3. Name two methods of determining inspection priorities.
4. List three hazards associated with holiday gatherings in large public assembly occupancies.
5. What is the primary hazard associated with institutional occupancies?
6. Do the model fire codes apply to residential properties?
7. What action is generally required by the fire official before attempting to secure an inspection warrant?

## DISCUSSION QUESTIONS

1. The mayor has requested that the city manager form a task force to perform "surgical enforcement" of city codes within an area of the city with a large ethnic population. The mayor believes that "flexing our muscle" will help stop potential problems that he fears may arise in the future because of the groups' "propensity to cause trouble due to their morally bankrupt culture." The rate of fire incidents and fire loss is below the city average. What are the problems with the mayor's plan?
2. Consider the foregoing scenario except that 70 percent of the fires and 90 percent of the fire fatalities in the city occur within the area. Should the fire prevention bureau target the area? What type of programs would you implement?

## CHAPTER PROJECTS

Using fire investigation reports from the U.S. Fire Administration, NFPA, or a source of your choice, identify the causes and contributing factors in three large-loss fires. Write a summary of each and explain what, if any, effect an inspection program might have had on the fire. If a program was in place, identify its impact (or lack thereof) and what might have been done to make the program more effective.

Using the National Fire Academy Learning Resource Center (http://www.lrc.fema .gov), research EFO papers and describe methodologies used to justify and measure the effectiveness of fire inspections programs. Describe which programs appeared to have the greatest and the least impact on life safety and property protection.

## ADDITIONAL RESOURCES

In-depth information on many of the subjects discussed in this chapter can be found in the following texts and publications and at these Web sites.

David Diamantes, *Fire Code Training and Consulting* at www.efirecode.com.

David Diamantes, *Fire Prevention: Inspection and Code Enforcement*, 3rd ed. (Delmar, 2006).

David Diamantes, "The Station Fire—Now We Remember the Cocoanut Grove," *The Fire Rattle*, 2003. Available at www.efirecode.com

John R. Hall, *Fire Code Inspections and Fire Prevention: What Methods Lead to Success?* (National Fire Protection Association, 1978).

## NOTES

1. John R. Hall, et al., *Fire Code Inspections and Fire Prevention: What Methods Lead to Success?* (Boston, MA: National Fire Protection Association, 1978), page viii.

2. "The Value and Purpose of Fire Department Inspections," *Special Interest Bulletin No. 5* (New York: American Insurance Association, 1975), page 1.

3. *2006 International Fire Code Commentary* (Falls Church, VA: International Code Council, 2006), page 577.

4. *BOCA Basic Fire Prevention Code*, 1981 ed. (Homewood, IL: Building Officials and Code Administrators, 1981), page 57.

5. Ibid., page 5.

6. *International Building Code Commentary*—Volume 1, 2006 ed. (Falls Church, VA: International Code Council, 2006), pages 4–63.

7. *Jacobellis v. Ohio*, 378 U.S. 184, 197 (1964).

8. Michael J. Karter, Jr. "Fire in the United States," *NFPA Journal*, September/October 2009.

9. Larry D. Lamb, "Rental Inspections Under Attack in Michigan. An Examination of Landlord and Tenant Opinions in Niles, Michigan." Emmitsburg, MD: National Fire Academy. November 2005.

# 7

# Fire Protection Systems Testing

## LEARNING OBJECTIVES

Upon completion of this chapter, you should be able to:

● Describe the fire code official's role in the inspection and testing of fire protection systems.

● Explain the importance of systems acceptance tests, maintenance and periodic inspections, and retests.

● List four system elements that contribute to reliability.

● Discuss the impact of unreliable fire protection systems on the public, the business community, and the fire service.

# ACTIVE FIRE PROTECTION

Within the trade, the term **active fire protection feature** generally means fire extinguishing and smoke control and alarm systems versus **passive fire protection** features, which means fire resistance–rated construction. Active fire protection features have moving parts—they do things or perform some type of operation. Passive fire protection features, such as structurally applied coatings of concrete or gypsum that shield a building's skeleton from flames and heat or rated walls that separate occupancies, are always there and require no activation or signal. There is less to go wrong with the passive features, and they require less maintenance and less vigilance to ensure proper operation.

**active fire protection feature**  a fire protection feature or system such as automatic sprinklers, fire detectors, or smoke removal systems that operate or activate automatically or manually.

**passive fire protection**  built-in fire protection features such as rated construction that provide fire safety and do not require activation.

If you were to ask the assembled fire chiefs from across the United States if they support the installation of active fire protection features such as sprinkler systems and alarm systems, you would undoubtedly get a "Yes," perhaps qualified with "if the systems are properly installed and maintained." Properly installed systems have a huge potential impact for safety and property protection. Improperly designed, installed, or maintained fire protection systems can pose problems for the fire department, the building owner or occupant, and the public. Problems include false or nuisance alarms that require fire department response, a false sense of security for building owners who think their buildings are protected by inoperative systems, a public that has been immunized by faulty alarms into thinking that every alarm signal is false, and a few cases in which inadequately designed or protected systems damage the very building they were installed to protect because of freezing or malfunction. None of these problems is minor.

**NOTE** Improperly designed, installed, or maintained fire protection systems can pose problems for the fire department, the building owner or occupant, and the public.

Firefighters and the public are put at risk by emergency vehicle responses to fire protection system activations. Increased response times result because units from other districts must be used for successive calls for service to the area. The general public's perception that every alarm signal must be a false alarm has resulted in genuine public apathy at best and significant public danger at worst in cases in which people have refused to evacuate structures involved in fire.

**NOTE** The general public's perception that every alarm signal must be a false alarm has resulted in genuine public apathy at best and significant public danger at worst.

# FIRE PROTECTION SYSTEMS

Active fire protection systems in buildings date back to early efforts by insurance underwriters to minimize the impact of fire on insured commercial properties. The early installations were rarely in response to building or fire code requirements. Most were the result of insurance company requirements or were installed to qualify for reduced premiums. Even the early model building codes required far fewer active fire protection features than today's codes. The 1940 *Uniform Building Code* required sprinklers only in large cellars of commercial buildings lacking windows above grade, the stage areas of assembly occupancies, woodworking facilities, the manufacture and renovation of mattresses, and cellulose nitrate motion picture film exchanges.[1] The only alarm component mentioned in the code was for required sprinkler systems.

Time has shown active fire protection features, particularly the automatic sprinkler system, to be of great value in reducing property loss, minimizing business interruption, and protecting the lives of building occupants. Active fire protection systems combined with adequately designed and maintained means of egress and fire resistance–rated construction provide the greatest level of protection to the public, to the economic vitality of the businesses they protect, and to the safety of firefighters who are called on to extinguish structure fires. Effectiveness depends on reliability. In describing the importance of reliability in fire alarm systems, John M. Cholin, P.E., stated that reliability depends on the contribution of four system elements: design, equipment, installation, and maintenance.[2] The same elements are applicable to all active fire protection features, including extinguishing systems, smoke control systems, and alarms. The fire service has a vested interest in the reliability of these systems.

**NOTE** Effectiveness depends on reliability.

# WHICH BUILDINGS AND WHY

If you spend enough time searching high and low, you will find buildings that are equipped with sprinkler systems that are neither required by code nor were installed in order to get one of the **trade-off incentives** contained in the model building codes. The term *trade-off* has a sordid connotation to some people, and many in the fire service believe the idea is less than wholesome. However, the term describes provisions within model construction codes that permit the reduction of certain requirements (generally passive fire protection features) in return for the installation of sprinklers.

**trade-off incentive** code provisions that permit the reduction of fire protection ratings and increases in building height, area, number of stories, and reduction in means of egress provisions in return for the installation of sprinkler systems.

Building codes have always permitted the reduction of certain fire resistance–rated features in sprinklered buildings. The basic idea makes good sense because the odds that a major fire will occur in a building fully protected with a properly maintained sprinkler system is extremely low. Records that date back well over a century bear out the fact that automatic sprinklers are the most cost-effective and efficient method of protecting property from fire that has ever existed. The chances of dying in a fire, as well as the average property loss per fire, are cut by one-half to two-thirds in sprinklered buildings.[3]

**NOTE** The chances of dying in a fire, as well as the average property loss per fire, are cut by one-half to two-thirds in sprinklered buildings.

Figure 7-1 summarizes the reductions in property loss associated with sprinklers as reported by Kimberly D. Rohr in *U.S. Experience with Sprinklers.*[4]

The installation of sprinkler systems is of such benefit to building owners, occupants, and to the fire service that strategies to increase the number of sprinkler systems in buildings have been developed and implemented by fire service organizations. Fire service demands that model codes require sprinklers in more buildings were often challenged and defeated in code hearings by special interests. As a result, methods of inducing developers and owners to install sprinklers were embraced. Reducing the costs of other fire safety features, thereby making the installation of sprinklers cost effective, was the trade-off. It was, and still is, merely a system of balancing active and passive fire protection features.

In return for a complete sprinkler system, a building can be an additional story in height, the building can be larger in size, and fire resistance ratings of structural members can be reduced (Figure 7-2). Travel distance to exits can be increased based on the knowledge that sprinkler systems greatly reduce the rate of fire growth and the probability of flashover. The fire service did not invent the incentive system; rather, it embraced it as a method of increasing the likelihood of sprinkler protection. The number of sprinkler systems installed in buildings has increased considerably because of the economic incentives within the model construction codes. The idea of trade-offs is a good one. However, the question that must be constantly asked is just how much can be safely traded away?

**FIGURE 7-1**

Comparison of the average loss per fire based on sprinkler protection and the loss reduction afforded by sprinklers.

| Occupancy Type | Buildings without Sprinklers | Buildings with Sprinklers | Fire Loss Reduction (%) |
|---|---|---|---|
| Stores and Offices | $25,000 | $11,700 | 53 |
| Manufacturing | $52,500 | $18,700 | 64 |
| Health Care | $ 4,800 | $ 1,700 | 66 |
| Public Assembly | $21,800 | $ 6,500 | 70 |

**FIGURE 7-2**
The installation of a sprinkler system allows these apartment buildings to be four stories in height and constructed with combustible materials.

**NOTE** The idea of trade-offs is a good one. However, the question that must be constantly asked is just how much can be safely traded away?

An in-depth discussion of the building code system and of balancing active and passive fire protection features is beyond the scope of this text. However, the building code change system's existence and fire service involvement in the code process is within the scope. Trade-offs exist. Building structural elements are less fire resistant, combustible buildings are bigger and taller, and the ability for occupants to rapidly escape is reduced in sprinklered buildings. Someone needs to ensure that the systems are properly installed and maintained. That someone should be the local fire prevention bureau.

Automatic sprinklers and other fire protection systems are required in buildings and structures that contain high-risk occupants such as hospitals, prisons, and schools, as well as in buildings with large occupant loads such as theaters, nightclubs, and passenger terminals. They are also required where types and amounts of regulated materials are present and in buildings with high fire loads such as warehouses and stores. In most cases, these buildings receive the benefit of reductions in passive fire protection even though they have to be sprinklered anyway.

Low-risk structures such as office buildings and small residential and industrial buildings that rarely require sprinklers based on occupancy classification are equipped with sprinklers when the owner realizes that savings from reductions in passive fire

**FIGURE 7-3**

Examples of model building code incentives for the installation of fire sprinklers. (From the *2000 International Burning Code.*)

| BENEFIT OR INCENTIVE | EXAMPLE |
|---|---|
| Height increase | Nonsprinklered, unprotected frame apartment buildings are limited to two stories, three stories if sprinklered. |
| Area increase/Reduction in construction class | Building areas based on construction type may be increased by 300 percent for one-story buildings, 200 percent for multistory buildings. |
| Reduction in structural ratings | Protection of structural members through encasement in gypsum, concrete, or other material may be reduced. |
| Reduction in occupancy separation ratings | Ratings for walls and floor/ceilings between different uses may be reduced. |
| Increased travel distance to exits | Distance to an exit may be increased 50–100 feet in some occupancies. |
| Decrease in separation of exits | Exits may be closer together, increasing the chance that they could become impassible from the same cause. |
| Reduction in requirements for interior finish/decorative materials | Flamespread ratings are reduced in sprinklered buildings and increased amounts of decorative materials are permitted in sprinklered buildings. |

protection features are greater than the cost of sprinkler installation. Figure 7-3 contains some of the code incentives for sprinkler system installation. It is by no means all-inclusive.

## Sprinkler Reliability

Sprinkler systems are very reliable. In a 1985 *Canadian Building Digest* article, J.K. Richardson placed the reliability of sprinkler systems at greater than 96 percent based on statistics compiled by the National Fire Protection Association (NFPA), the Australian Fire Protection Association, and the city of New York. The statistics were maintained from the late 1800s through the 1960s.[5]

**NOTE** Sprinkler systems are very reliable.

The reliability of fire protection systems depends on proper design, installation, and maintenance. The safety of building occupants and viability of the businesses that are protected are at stake. The reliability of sprinklers becomes even more of an issue when the code-permitted reductions listed in Figure 7-3 are considered. The NFPA

✳ **FIGURE 7-4**

The leading causes of unsatisfactory sprinkler performance 1925–1969. (Source: Data from Kimberly D. Rohr, *U.S. Experience with Sprinklers*, National Fire Protection Association, Quincy, MA, 2001, page 50.)

| PROBLEM | PERCENTAGE OF CASES |
|---|---|
| Water shutoff | 35.4 |
| System inadequate for level of hazard | 13.5 |
| Inadequate water supply | 9.9 |
| Inadequate maintenance | 8.4 |
| Obstruction to water system | 8.2 |
| System only designed for partial protection | 8.1 |

discontinued tracking the performance of sprinklers in 1970, citing that the data collection was biased toward poor performance, which could result in an inaccurate perception of sprinkler effectiveness.[6] Nonetheless, information from statistics collected before 1970 clearly identifies areas that require heightened vigilance by owners, design professionals, contractors, and code officials. Figure 7-4 lists the top six causes cited for unsatisfactory sprinkler performance from 1925 to 1969.[7]

**NOTE** The reliability of fire protection systems depends on proper design, installation, and maintenance.

The primary causes for unsatisfactory performance before 1970 were issues that could be readily identified by inspection and testing. As far as keeping statistics goes, it also shows that if you do not like what you see, stop watching. In 1970, most people in fire protection never envisioned the sprinkler recalls that began to occur in the 1990s. The beauty of sprinklers was their simplicity, some of which faded with the development of high-performance heads.

The old insurance engineers were sticklers that sprinklers be correctly designed and installed, and they went to great pains to minimize the possibility that sprinkler systems could be rendered inoperable without considerable effort. Standards were written with reliability in mind. **Outside screw and yoke valves** (Figure 7-5) and post indicator valves (Figure 7-6) for sprinkler water supplies were designed to be instantly recognizable as open or shut. The term "closed" was not used on post indicator valves because closed and open do not look different enough to ensure that the valves could be read from a distance. Although the systems were complex, reliability was a foremost concern.

**outside screw and yoke valves** a fire protection system water supply valve in which the stem protrudes from the housing when the valve is shut, making the valve condition readily apparent.

**FIGURE 7-5**
Closed valves are the leading cause of sprinkler failure.

**FIGURE 7-6**
Post indicator valve shows that valve is shut rather than closed.

## Fire Departments and Fire Protection System Inspections

The involvement of fire departments in the inspection of new fire protection systems is a fairly recent development. In describing "fire department inspection work," the 1941 *Crosby-Fiske-Forster Handbook of Fire Protection*, which later became the *NFPA Fire Protection Handbook*, mentioned only "enforcing the maintenance of automatic sprinklers, standpipes, hose and extinguishers."[8] The inspection and approval of new fire protection systems was—and in many jurisdictions still is—the responsibility of the building code official. In some cases, fire departments are forced to live with what is approved by the building department.

**NOTE** In some cases, fire departments are forced to live with what is approved by the building department.

# FIRE PROTECTION SYSTEMS TESTS

## Fire Codes and Acceptance Tests

The model fire codes have mandated acceptance tests and contractor's certification of completion for all required fire suppression and alarm systems since the 1970s. The *BOCA/National Fire Prevention Code* (published by the Building Officials and Code Administrators) specifically mandated witnessing of the tests by the fire official, and the *Standard Fire Code* and *Uniform Fire Code* required notification of the fire official before the tests, thus enabling the fire official to witness the tests. The *International Fire Code*, the cooperative effort of the members of all three model code organizations, requires a contractor's "statement of compliance," prior notification of the fire code official, and the fire code official's approval of acceptance tests, thus enabling fire departments to engage in on-scene inspections during acceptance testing.

Required acceptance tests and test protocols are included within the code-referenced standards for installation. The NFPA standards for fire protection systems shown in Figure 7-7 are among those referenced by the *International Building Code* and *International Fire Code*.

**FIGURE 7-7**

NFPA maintains standards for fire protection systems.

| STANDARD NUMBER | TITLE |
|---|---|
| NFPA 11 | Low-, Medium-, and High-Expansion Foam |
| NFPA 11A | Medium- and High-Expansion Foam Systems |
| NFPA 12 | Carbon Dioxide Extinguishing Systems |
| NFPA 13 | Installation of Sprinkler Systems |
| NFPA 13D | Installation of Sprinkler Systems in One- and Two-Family Dwellings and Manufactured Homes |
| NFPA 13R | Installation of Sprinkler Systems in Residential Occupancies up to and Including Four Stories in Height |
| NFPA 14 | Installation of Standpipe and Hose Systems |
| NFPA 15 | Water Spray Fixed Systems for Fire Protection |
| NFPA 16 | Installation of Foam-Water Sprinkler and Foam-Water Spray Systems |
| NFPA 17 | Dry Chemical Extinguishing Systems |
| NFPA 17A | Wet Chemical Extinguishing Systems |
| NFPA 20 | Installation of Stationary Pumps for Fire Protection |
| NFPA 72® | National Fire Alarm Code® |

## Systems Testing Benefits

Why should the municipal fire department choose to dedicate its limited staff resources to the hands-on review of fire protection plans and the witnessing of required performance or acceptance tests required by the installation standards? Isn't it enough that another government agency or a licensed and bonded **third-party agency** ensure that systems are properly designed, installed, and maintained?

> **third-party agency**  nongovernmental agency that has been approved by the code official to conduct tests and inspections; test and inspection results are forwarded to a code official, who approves or rejects the installation, construction, or operation.

Each of the standards for installation referenced by the model building and fire codes include detailed provisions for the installation, testing, and maintenance of all active fire protection systems. The systems must undergo an in-depth test to verify that all components are properly installed and that they operate in accordance with their listing documents and the referenced standards. Physical performance of the test and conformance with the code are the responsibility of the installing contractor. The contractor is also required to notify the appropriate code official for the jurisdiction before the final acceptance test is performed. The code official has the option of witnessing the acceptance test. The time spent in this endeavor is time well spent. Witnessing of acceptance tests by fire department personnel:

- Ensures that systems are installed in accordance with the approved plans.
- Ensures that conditions in the structure, including life and property hazards, accurately reflect those used to design the system. Fire protection systems are engineered systems designed for specific hazards.
- Provides an additional set of eyes on the construction site, promoting construction safety.
- Creates opportunities for interaction between fire department and building department personnel, leading to increased cooperation and safer structures.
- Enables fire department inspectors and contractors to interact, leading to a better understanding of both group's expectations and needs.
- Provides valuable training for firefighters and prospective fire officers in the design and operation of fire protection systems—training that unfortunately is scarce within the framework of most local fire department training academies (Figure 7-8).

## Maintenance and Retests

Ensuring that systems are properly designed and installed accomplishes little if they are not adequately maintained and repaired. System maintenance and retests are required by many of the same standards as are used for installation (Figure 7-9). One exception is for sprinklers

**FIGURE 7-8**
Inspectors who witness fire protection system tests quickly become a resource for the entire fire department. (Photo courtesy of Maurice Jones.)

**FIGURE 7-9**
NFPA 25 requires flow tests of specific water-based fire protection systems.

and other water-based fire protection. *NFPA 25: Standard for the Inspection, Testing, and Maintenance of Water-Based Fire Protection Systems* evolved from *NFPA 13A: Recommended Practice for Inspection, Testing and Maintenance of Sprinkler Systems* and *NFPA 14: Recommended Practice for Inspection, Testing and Maintenance of Standpipe Systems.*

**NOTE** Ensuring that systems are properly designed and installed accomplishes little if they are not adequately maintained and repaired.

"Recommended practices" are exactly what the name implies—recommendations, and as such, they are not legally enforceable. There was a demonstrated need to provide a user-friendly document that includes nearly all water-based fire protection systems. (Sprinkler systems in one-and two-family dwellings and manufactured homes are exempted from *NFPA 25*.) *NFPA 25* includes maintenance provisions; tests protocols and schedules; and even copies of useful checklists that summarize weekly, monthly, and annual tests. Test records must be available on-site for inspectors to examine (Figure 7-10). Retests and inspections required by *NFPA 25* are conducted by plant personnel or outside contractors. The fire prevention bureau ensures that the tests and maintenance are being performed.

What sometimes gets lost in the shuffle is that maintenance and testing of these complex systems not only ensures safety and continuity of business operations but also saves the owner money in the long run. Maintenance is cheaper than extensive repairs or replacement.

**FIGURE 7-10**
Contractor's record and tags are evidence of compliance with code-mandated maintenance and tests. (Courtesy of Tom Herman, Eagle Fire Protection.)

**NOTE** Maintenance is cheaper than extensive repairs or replacement.

## Contractor Oversight

As in any regulatory process, the job of the fire inspector in the testing and inspection of fire protection systems is to ensure that the contractor performs the job in accordance with the code, not to tell the contractor how to perform the job. The inspector is responsible for ensuring that the code provisions are met and that the owner and the public will be safer because of these efforts. This is not a game of "gotcha," and the inspector should not get involved in contract disputes between the owner and contractor or between the general contractor and the subcontractors.

**NOTE** The job of the fire inspector in the testing and inspection of fire protection systems is to ensure that the contractor performs the job in accordance with the code, not to tell the contractor how to perform the job.

Requirements for plan submissions and permits, test protocols and fees, and scheduling procedures should be clearly spelled out, readily available, and fairly implemented. Playing favorites is not only unethical; it is illegal. Allowing one contractor to skirt the rules gives him or her an unfair advantage in a highly competitive trade. The old axioms of "fair, firm, and flexible" should describe the inspection and test program.

Keep in mind that construction is a fast-paced, highly competitive business. Developers attempt to work with detailed budgets, financing the total costs, including inspection and test fees. After construction financing is obtained, unexpected costs and change orders result in budget overruns. These costs must be paid in cash by the developers, and they cut directly into profit margins that are often thin to start. Most developers want timely inspection service, want to know what is expected of them, and want to know the cost at the start of the project. Most are willing to pay fees that allow the jurisdiction to provide timely inspection. They want to be able to budget for the costs and want what they are paying for—timely inspections. Delays in inspections and approvals slow down construction. Many construction contracts contain hefty bonuses for on-time completion and even heftier fines for delays.

## DEVELOPING AN ACCEPTANCE TESTING PROGRAM

The following questions must be answered before contemplating a program for the inspection and witnessing of acceptance tests:

- What agency has the oversight responsibility for fire protection systems installation within the local code? The fact that the local building department is responsible does not necessarily preclude the fire department from functioning as an agent of the building official.

● How many and what type of systems are installed each year? Is the area prone to construction booms and busts? Could staff be hired and trained only to have the program dry up because of lack of work? Such instances may point toward fire department staff using contracted engineers and overseeing the overseers.

● Is the fire department willing to provide the necessary personnel, training, and resources to adequately perform the job? Staff or resource shortages are not acceptable excuses for long delays in providing inspection services. The potential damage posed by doing a less-than-adequate job is very real and very lasting on the image of the fire department.

The National Fire Academy in Emmitsburg, Maryland, is a good source for information on fire protection systems testing programs, as well as being the premier location for training on fire protection systems, systems testing, and plan review. The NFPA and the International Code Council also provide training and certification programs.

## SUMMARY

Active fire protection systems include extinguishing, smoke control, and alarm systems. These complex engineered systems have a well-documented record of increasing the level of life safety and minimizing property damage and business interruption when the systems are properly designed, installed, and maintained. The systems must function with a high degree of reliability or they can actually negatively impact public safety and the preservation of property through false function. Reliability depends on four system elements: design, equipment, installation, and maintenance. Modern fire departments have a vested interest in the reliability of these systems.

## REVIEW QUESTIONS

1. List two active fire protection systems for buildings.
2. List two passive fire protection systems for buildings.
3. What four system elements determine reliability?
4. What NFPA standard was developed for the testing and maintenance of water-based fire protection systems?
5. Who is responsible for physically performing acceptance tests of fire protection systems?

## DISCUSSION QUESTIONS

1. The NFPA discontinued tracking the performance of sprinklers in 1970, citing that the data collection was biased toward poor performance. Do you see value in gathering, compiling and publishing these data?
2. Write a short position paper either supporting the NFPA's position or opposing it. Use fire data or information from actual fire incidents or perform your own limited survey of sprinkler systems.

## CHAPTER PROJECT

Your local contractors' association has asked that a representative from the fire department attend its next business meeting and give a presentation regarding the value of inspections and witnessing acceptance tests by the fire prevention bureau. Develop an outline for the presentation.

## ADDITIONAL RESOURCES

In-depth information on many of the subjects discussed in this chapter can be found in the following texts and publications and at these Web sites.

John L. Bryan, *Automatic Sprinkler and Standpipe Systems* (National Fire Protection Association, 1990).

Arthur Cote and Percy Bugbee, *Principles of Fire Protection* (National Fire Protection Association, 1988).

David Diamantes, *Fire Prevention: Inspection and Code Enforcement*, 2nd ed. (Delmar, 2003).

Robert M. Gagnon, *Design of Special Hazard and Fire Alarm Systems* (Delmar, 1997).

Robert M. Gagnon, *Design of Water-Based Fire Protection Systems* (Delmar, 1997).

Maurice Jones, *Fire Protection Systems* (Delmar/Cengage, 2009).

National Fire Protection Association, *Fire Protection Handbook* (National Fire Protection Association, 2008 ed.).

National Fire Protection Association at www.nfpa.org.

Kimberly D. Rohr, *U.S. Experience with Sprinklers* (National Fire Protection Association, 2001).

U.S. Fire Administration at www.usfa.fema.gov.

## NOTES

1. *Uniform Building Code*, 1940 ed. (Los Angeles: Pacific Coast Building Officials' Conference, 1940), page 246.
2. Arthur E. Cote, P.E., et al., *Fire Protection Handbook*, 18th ed. (Quincy, MA: National Fire Protection Association, 1997), pages 5–37.
3. Kimberly D. Rohr, *U.S. Experience with Sprinklers* (Quincy MA: National Fire Protection Association, 2001), page 1.
4. Ibid.
5. J.K. Richardson, "The Reliability of Automatic Sprinkler Systems," *Canadian Building Digest*, vol. 238, July 1985.
6. Rohr, page 48.
7. Ibid., page 50.
8. Robert S. Moulton, et al., *Crosby-Fiske-Forster Handbook of Fire Protection* (Boston: National Fire Protection Association, 1941), page 539.

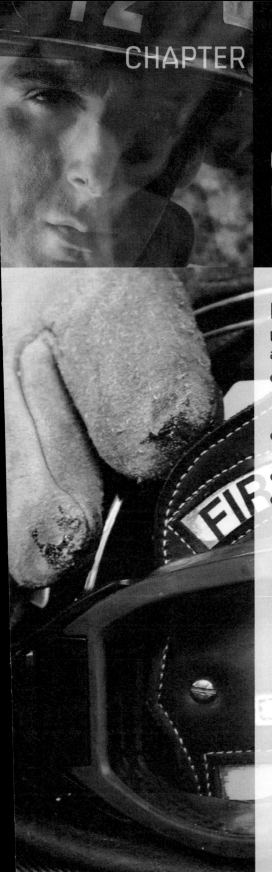

# 8

# Other Fire Prevention Functions

## LEARNING OBJECTIVES

Upon completion of this chapter, you should be able to:

- List nontraditional government functions that are sometimes assigned to fire prevention bureaus.
- Describe the rationale used to justify this action.
- Discuss the potential benefits for the public.
- Describe the potential negative impact on the organization's mission of fire prevention.

# FIRE PREVENTION–RELATED FUNCTIONS

There is an elastic quality that fire departments in the United States have historically possessed. It probably originates in the very nature of the organization. What other group can mobilize as quickly and respond at breakneck speed to almost any emergency situation? Whether it is a child trapped in a chimney, a corrosive liquid spill, or a city block–sized lumberyard fire, most communities expect the fire department to come and make it better. It is only natural that fire prevention bureaus have branched into functions that sometimes complement and often enhance their fire prevention and protection efforts.

Some bureaus are tasked with functions that appear to have little to do with life safety and property protection. The danger here is that the original mission can be superseded by tasks that are perhaps more glamorous and more news friendly. Ensuring that child safety seats are properly installed in automobiles is a worthy cause; any program aimed at protecting the lives of the public is a worthy one.

> **NOTE** The danger here is that the original mission can be superseded by tasks that are perhaps more glamorous and more news friendly.

Do not let the life safety mantra totally obscure the issue of property protection. If life safety is the only measure of success for a fire prevention bureau, consider this: The Centers for Disease Control and Prevention reported 3,852 victims of drowning in 2005.[1] For the same year, the the National Fire Protection Association reported 3,675 fire fatalities.[2] The difference between the two is statistically insignificant, but more people drowned than were killed by fire. Should we throw away our codebooks, don swimsuits, and hang around the pool looking for swimmers in distress? Aren't drowning victims just as important as fire victims? Instead of "change your clock; check your smoke alarm," should we add "learn to swim" to the slogan and preach universal swimming instruction for all Americans?

Better yet, the National Safety Council report listed 4,600 choking fatalities the same year.[3] If we combine swimming lessons with a National "Chew Your Food" campaign, we could really have an impact on the number of fatal accidents. But then, we would be forgetting that we have this fire problem. "America today has the highest fire losses in terms of both frequency and total losses of any modern technological society,"[4] reported *America at Risk* in 2000. Our mission extends beyond life safety. Fire threatens the economic viability of every community. When people lose their jobs, the impact is far greater than just the loss of disposable income. Medical care for their children is sometimes deferred. Mothers put off routine medical care such as mammograms. People get sick. Families are uprooted.

> **NOTE** Fire threatens the economic viability of every community.

Some government functions can be neatly dovetailed into the fire prevention bureau's traditional mission and actually enhance the fire prevention mission. Enforcement of construction and property maintenance codes has successfully been integrated within some fire departments in the United States. It takes a strong, confident fire chief who can see the big picture and who is willing to rely on staff members with technical skills vastly different from those of most firefighters.

For some localities, the investigation of hazardous waste and environmental crimes was the natural extension of materials regulation, emergency incident mitigation, and fire investigation. What other agency has the knowledge base and operational experience to handle such complex cases? Additionally, fire departments and fire prevention bureaus have a vested interest. Firefighters respond to every fire, explosion, and hazardous materials release. Who else cares as much?

## CONSTRUCTION REGULATION

Construction regulation requires technical competence in structural, mechanical, electrical, and plumbing engineering and the associated trades. Model codes contain specific requirements for construction officials and their assistants. Certification is sometimes mandated but always desirable. **Professional registration** or licensure in the fields of engineering or architecture is also highly desirable (Figure 8-1).

| professional registration | state required licensure system for professionals such as engineers and architects. |
|---|---|

Few fire chiefs or chief fire marshals have the education, the training, or even the desire to become professional engineers or architects. But they do not need to be engineers or architects to manage construction regulation as part of the department's overall mission of life safety and property protection (Figure 8-2). Within fire departments where construction regulation has been successfully integrated, the title of building official and its regulatory responsibilities belong to a manager who reports to the fire chief or fire marshal. Chief mechanical, plumbing, and electrical inspectors and their staffs report to the building official.

**FIGURE 8-1**
Fire departments that regulate construction typically employ civilian engineers or architects.

**FIGURE 8-2**
Fire departments
have a vested
interest in regulating
construction
as evidenced
by this blocked
fire department
connection at a
construction site.
(Photo courtesy of
Duane Perry.)

The Bureau of Construction, Office of Code Enforcement, Office of Building Regulation, or whatever name the organization uses is a division within the fire department or within the fire prevention bureau. The important part is not the name; it is the function. That one agency is now in charge of every building and structure from the time it starts as a blueprint, through construction, renovation, and finally destruction by demolition or fire.

Some cities have consolidated all inspection functions within one agency, other than the fire department, in an attempt to reduce duplication. The notion that all inspections are basically alike and that cross-training can overcome all, is too simplistic. Streamlining government or reduction of duplication is not the reason jurisdictions have transferred the building official and associated staff to the fire department. When construction regulation falls within the mission of the fire department, it is rightly recognized for what it really is—a public safety function.

**NOTE** The notion that all inspections are basically alike and that cross-training can overcome all is too simplistic.

Fire departments operate 24 hours a day, every day, and have resources and infrastructure that can enhance construction regulation. When construction engineers and inspectors are part of the fire department, both cultures mesh, expanding the knowledge base and operational capabilities of both groups. Fire departments benefit from the technical knowledge of structural engineers; engineers benefit from the knowledge of fire behavior and fire department tactics that can only be gained from direct personal contact.

# PROPERTY MAINTENANCE CODE ENFORCEMENT

Property maintenance codes provide minimum standards for existing structures for light, ventilation, space, heating, sanitation, protection from the weather, life safety, and fire safety. Property maintenance codes address both indoor and outdoor conditions and are used by many urban and suburban jurisdictions to ensure that properties are maintained and to prevent blighted neighborhoods. They are the only model codes to specifically require a specific amount of living space per occupant. Fire code occupant loads and means of egress requirements strictly address exiting as a safety issue.

Although some might claim that property maintenance codes go well beyond the prevention of fire, few would argue that many of the code provisions are directly involved with fire safety or have an impact on fire safety. Faulty or improperly maintained heating, lighting, ventilation, and electrical distribution equipment are major fire causes. When landlords provide substandard central heating systems, tenants are forced to rely on space heaters. In 1998, space heaters accounted for 65 percent of the home heating fires and 76 percent of the fatalities associated with them.[5] From 1994 through 1998, electrical distribution equipment was the fifth leading cause of home structure fires, fourth in home fire deaths, seventh in home fire injuries, and second in direct property damage.[6]

Property maintenance codes are typically adopted in concert with ordinances that require the registration and inspection of residential rental properties. These ordinances typically require inspection with each change of tenancy. Some ordinances mandate inspections at 1- or 2-year intervals. Because 70 to 75 percent of civilian fire deaths and injuries occur in residences, property maintenance code inspections have significant potential to reduce the rate of deaths and injuries, especially in jurisdictions with large stocks of rental properties.[7]

**NOTE** Because 70 to 75 percent of civilian fire deaths and injuries occur in residences, property maintenance code inspections have significant potential to reduce the rate of deaths and injuries, especially in jurisdictions with large stocks of rental properties.

## Political Considerations

Public perception of property maintenance code inspection programs is highly influenced by media reports and the views of neighborhood leaders. Frequent contact among code officials and church, civic, and social groups within the community can do much to dispel concerns that programs are intrusive and conducted for reasons other than property maintenance. Inspectors are legally obligated to report illegal activity encountered during inspections. Using the inspection process as a pretext to gain entry for other purposes is illegal and sure to destroy relationships with community leaders and eliminate the good will of the tenants.

**NOTE** Inspectors are legally obligated to report illegal activity encountered during inspections.

## Vacant Structures

One of the most important objects of property maintenance codes is securing vacant or abandoned properties (Figure 8-3). These properties become targets for vandalism, havens for illegal activity, and temporary shelters for homeless people. Vacant and abandoned properties are frequently involved in fires and have been the scenes of numerous multiple firefighter fatalities. On December 4, 1999, six Worcester, Massachusetts, firefighters were killed at a fire in an abandoned cold storage warehouse. The fire started when a homeless couple knocked over a candle. The firefighters died attempting to rescue the homeless couple they thought were trapped inside.

**NOTE** Vacant and abandoned properties are frequently involved in fire and have been the scenes of numerous multiple firefighter fatalities.

No other organization has the vested interest in securing vacant and abandoned buildings. The frequency with which abandoned structures have become havens for homeless people has created a crisis for the fire service. All would agree that abandoned structures do not justify placing firefighters at risk. When reports from bystanders indicate that the structures are occupied, incident commanders are placed in a no-win situation. A proactive campaign to reduce or eliminate these structures is the only acceptable response, whether by means of barricading or through demolition, with demolition being the preferred course of action.

---

**FIGURE 8-3**
No other organization has a greater vested interest in securing vacant structures than the fire department.

**NOTE** A proactive campaign to reduce or eliminate these structures is the only acceptable response, whether by means of barricading or through demolition.

## HAZARDOUS WASTE AND ENVIRONMENTAL CRIME INVESTIGATION

When the term *hazardous materials* first appeared in fire codes in the 1980s and 1990s, some fire prevention bureaus scoffed at the idea of hazardous materials regulation. "Let the hazmat team do it" was an common comment, usually followed by expletives. Those comments were usually made before the code was opened, however. The hazardous materials section had previously been the materials handling section, and although there were some new provisions, the materials were the same as before.

Fire prevention bureaus have been regulating the storage, handling, and use of hazardous materials since there have been fire codes. Fire departments have been extinguishing fires and mitigating spills, leaks, and releases since before there were fire codes. No other organization has more experience with such a wide range of materials, incidents, or situations. In many jurisdictions, fire prevention bureaus were the logical organization for hazardous materials and hazardous waste regulation, beyond the scope of the fire code, and for the investigation of environmental crimes (Figure 8-4).

**NOTE** Fire prevention bureaus have been regulating the storage, handling, and use of hazardous materials since there have been fire codes.

**FIGURE 8-4**
In many jurisdictions, fire departments and fire prevention bureaus were the logical pick to investigate hazardous waste crimes.

Fire departments are one of the few organizations that can contain and neutralize a hazardous materials release and, without skipping a beat, collect evidence of any associated crime and identify and arrest the suspect. The sheer quantity of protective equipment, ongoing training, and staffing resources needed to safely manage these incidents preclude many traditional law enforcement agencies from performing these investigations without the assistance of the fire department. Hazardous waste crime scenes are not unlike fire scenes. Fire department standard operating procedures that provide for fire scene security and safety during fire investigations can be carried over to hazardous waste crime or environmental crime scenes.

## HOMELAND SECURITY AND CIVIL DEFENSE

Fire departments are among the first to respond to acts of domestic terrorism. Fire prevention bureaus have a key role in ensuring that emergency plans and security measures developed by other government agencies and business and industry do not conflict with emergency response procedures. Never has the need for emergency planning and drills been greater.

**NOTE** Fire departments are among the first to respond to acts of domestic terrorism.

Fire prevention bureaus have extensive experience with fire department access roads, emergency egress and ingress, and special locking arrangements and systems. The threat of terrorist attack does not reduce the need for traditional fire protection and emergency egress systems; it enhances it. Poorly planned and implemented security measures can do significantly more harm than good. When the Pentagon was attacked on September 11, 2001, the ability to rapidly evacuate was critical. The ability to strategically position fire apparatus could have been impacted if barriers had been installed without proper planning.

**NOTE** The threat of terrorist attack does not reduce the need for traditional fire protection and emergency egress systems; it enhances it.

Safety and security are complementary, not competing, goals. Formal communications between safety and security agencies must be established at the highest levels of the organizations to ensure effective communications and cooperation are maintained at all levels. Working groups with representatives from multiple agencies can assist business and industry in developing effective safety and security measures. Security managers sometimes hear conflicting recommendations when agencies with different missions are contacted individually. Group meetings with representatives of all the agencies in attendance can greatly reduce this possibility.

**NOTE** Safety and security are complementary, not competing, goals.

## SUMMARY

The resources and culture of fire departments and fire prevention bureaus make them the logical choices to provide services that are sometimes considered nontraditional. Construction regulation, hazardous materials regulation and investigation, and civil defense are just some examples. Care must be taken to ensure that the original mission of the fire prevention bureau—property protection and safety from fire—is not lost in the process.

## REVIEW QUESTIONS

1. When applied to design professionals, what is meant by *professional registration*?
2. List five areas typically regulated by property maintenance codes.
3. What is meant by the property maintenance code term *space*?
4. Before establishing a property maintenance inspection program, what can code officials do to allay public concerns?
5. What program is often adopted in concert with a property maintenance code?
6. Hazardous waste crime scenes have similar safety and security concerns as do what other type of scene?
7. List two issues of concern to departments regarding plans to harden buildings against terrorist attack.

## DISCUSSION QUESTIONS

1. You receive a call from the security director for the local cable television provider. She has a list of addresses for properties suspected of pirating cable TV service and requests that inspectors enter the properties and perform property maintenance code inspections. Occupancies found to have illegal connections can be reported to her because pirating cable service is a criminal misdemeanor. Will you honor her request? Justify your answer.
2. A member of the city council has proposed that an "amnesty" be declared, prohibiting city agencies from enforcing any law or regulation that would "harm, harass, or force the removal" of homeless persons from vacant structures. During the last council meeting, the council member stated that homeless people are victims of an unfair capitalistic system and referred to city agencies as "jack-booted thugs," Prepare a short press release for the chief, addressing the fire prevention bureau's safety concerns.

## CHAPTER PROJECT

Prepare a position paper stating the role of the fire prevention bureau within your jurisdiction's homeland security plan or a theoretical model role in a jurisdiction of your choice.

## ADDITIONAL RESOURCES

In-depth information on many of the subjects discussed in this chapter can be found in the following texts and publications and at these Web sites.

Dennis Compton, et al., *Managing Fire and Rescue Services* (Washington, D.C., International City/County Managers Association, 2002).

Howard Dean, *Legal Aspects of Code Administration* (Washington, D.C.: International Code Council, 2002).

International Code Council at www.iccsafe.org.

National Fire Protection Association at www.nfpa.org.

Robert E. O'Bannon, et al., *Building Department Administration* (Washington, D.C: International Code Council, 2007).

U.S. Fire Administration at www.usfa.fema.gov.

## NOTES

1. Centers for Disease Control and Prevention, National Center for Injury Prevention and Control. *Web-based Injury Statistics Query and Reporting System (WISQARS)*, 2008. Available at www.cdc.gov/ncipc/wisqars.

2. Marty Ahrens, *Overview of the U.S. Fire Problem* (Boston MA: National Fire Protection Association, 2006).

3. *Act Fast to Stop Choking* (Washington, DC: National Safety Council, 2006).

4. *America at Risk* (Emmitsburg, MD: Recommissioned Panel for America Burning, Federal Emergency Management Agency, May 2000), page 15.

5. John R. Hall, *U.S. Home Heating Fires and Trends* (Quincy, MA: National Fire Protection Association, 2001), page iii.

6. Marty Ahrens, *Selections from the U.S. Fire Problem Overview Report* (Quincy, MA: National Fire Protection Association, 2001), page 56.

7. *Profile of Fire in the United States*, 12th ed. (Emmitsburg, MD: United States Fire Administration, National Fire Data Center, August 2001), page 3.

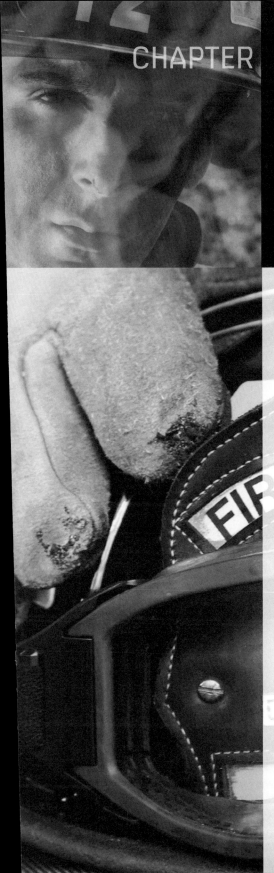

CHAPTER

# 9

# Fire Prevention Through Investigation

## LEARNING OBJECTIVES

Upon completion of this chapter, you should be able to:

- Identify local, state, and federal agencies involved in the investigation of fires.
- Describe the benefits of effective fire investigation.
- Identify the various roles adopted by fire departments to investigate fires.
- Identify agencies and organizations that provide fire investigations training.
- Discuss the use of case closure rate and conviction rate as measures of effectiveness for fire investigation units.

The commonly held misperception that fire destroys the evidence of a crime or that ashes and debris at the scene of a fire cannot be examined to develop solid information is wide-spread. However, countless convictions have resulted from the examination of the charred remains of arson fires. The meticulous dissection of electrical and mechanical equipment has led to the recall or repair of unsafe appliances and machinery. Government regulations for everything from flame-resistant treatments for carpets, children's sleepwear, and bedding are all rooted in the methodical examination of fire scenes. Fire code regulations regarding materials handling, processing, and storage, as well as emergency planning and preparedness, are all based on the fact that we found out how, where, and why fires start.

**NOTE** Countless convictions have resulted from the examination of the charred remains of arson fires.

## EARLY FIRE INVESTIGATION

Within 2 months of the Great Fire of London in 1666, a Frenchman described by Samuel Pepys as a "mopish, besotted fellow" was hanged for the crime of arson.[1] An investigation revealed that the fire had begun in the home of the king's baker, James Farynor. Kindling stacked near an oven ignited, and fire quickly spread through the house, forcing Farynor, his daughter, and servants to escape out an upper floor window. The fire burned for 5 days, destroying five-sixths of the area within the city walls and leaving nearly 200,000 people homeless.

During questioning, Farynor had taken great pains to give testimony to prove that the fire could not have been accidental. He had ample reason to fear being blamed for the conflagration. When a weak-minded foreigner publicly bragged of setting the fire, authorities were only too willing to arrest him. It was later discovered that the unfortunate man could not have started the fire. He had been aboard a Swedish ship and had not set foot ashore until 2 days after the fire. An account published 5 years after the fire squarely put the blame for the fire on the "drunkenness and supine negligence of the baker."[2] So much, for swift, sure justice.

Determining the origin and cause of fires is an important aspect in preventing them and in minimizing their destructive impact. If we are to prevent destructive fires and minimize their impact, we must know where, why, and how fires start and how fire behaves. By determining whether fires are accidental or the result of criminal acts, we are able to develop effective fire prevention programs and deter the crime of arson through criminal prosecution. In the wake of the Great Fire of London, the adoption of an ordinance regulating fuel storage might have accomplished more than a public hanging.

## THE REASON FOR FIRE INVESTIGATION

The Federal Bureau of Investigation (FBI)'s *Crime in the United States*, a compilation of Uniform Crime Reports (UCRs) submitted by law enforcement agencies, is prepared annually and distributed throughout the United States. Statistics are provided on case closure, namely those crimes in which someone is arrested, charged, and turned over to the courts for prosecution. Categories include murder, forcible rape, robbery, motor vehicle theft, burglary, and arson. Data are compared by geographical region, jurisdiction size (population), and jurisdiction type (cities or counties).

The report is used as a tool to gauge the effectiveness of crime prevention programs. A blanket statement that law enforcement agencies manage their response to crime solely by the report would be inaccurate. However, police chiefs, elected officials, and the media do use the report as a benchmark to compare their response to criminal activity with other jurisdictions. Sometimes the report makes great headlines.

The 2008 edition of *Crime in the United States* reported that crimes against people such as murder, aggravated assault, and rape had almost three times the closure rate of crimes against property, such as motor vehicle theft, burglary, or arson (45.1 percent and 17.4 percent, respectively). The differences were attributed to more extensive investigation efforts and the fact that crimes against people often have witnesses who can identify the perpetrator.[3] Of the 55,517 arson offenses reported for 2008, only 17.06 percent were closed by arrest, meaning those who commit the crime of arson have about the same statistical probability of arrest and prosecution as those who steal a motor vehicle.[4]

**NOTE** Those who commit the crime of arson have about the same statistical probability of arrest and prosecution as those who steal a motor vehicle.

The lesson here is not that police agencies do not care or do not take arson seriously. It is that they have many more violent crimes against people and must bring their resources to bear where they have the greatest potential effectiveness on the overall crime rate. The fire service has the greatest stake in the crime of arson. Not only does arson endanger the public; every fire also endangers the lives of firefighters. Who better then to take on the crime of arson?

**NOTE** The fire service has the greatest stake in the crime of arson.

## ORGANIZATIONS INVOLVED IN FIRE INVESTIGATION

An initial fire cause determination has been a traditional function of the fire incident commander. The unattended skillet left on the stove when the telephone rings begins a chain of events that most fire officers have seen many times. The origin and cause of cooking and other routine fires are generally determined by the officer in charge and noted on the incident report. Fire investigation by dedicated fire prevention bureau personnel; local, state, and federal police agencies; insurance companies; and private investigations firms have generally been limited to investigation of fatal fires, potential criminal acts, and large losses. Each organization has a mandate to fulfill, whether it is to serve the public, the stockholders of a corporation, or an individual with a vested interest in the incident.

For the most part, organizations, especially governmental agencies, are able to work cooperatively toward a common goal. At times, conflicting interests compete, adding one more element to an already complicated set of circumstances. It might be in the best interests of a particular person or business for a fire to be shown to have started in a particular manner. In the aftermath of the Great Fire of London, James Farynor was probably greatly relieved that another person had been convicted and hanged for starting the fire. The fact that the unfortunate man was mentally incompetent and was later found to have been aboard a ship were minor details to Farynor, to be sure. It is common for conflicting fire cause determinations to be presented by opposing legal counsel within the court system in both civil and criminal cases.

Instances of jurisdictional conflicts among federal, state, and local officials or between law enforcement and fire prevention bureau officials are an unfortunate, although infrequent, fact of life. The complexity of the process, dedication of staff and financial resources, and political implications, particularly involving law enforcement functions, have led some fire chiefs to steer clear of the function. However, no other organization has a greater vested interest in the investigation or prevention of fires than the fire department. It is our brothers and sisters who are put in harm's way on every single alarm.

## Investigation by the Fire Department

Fire departments participate in the investigation of fire to some extent in practically every jurisdiction. The volunteer chief of a rural department who suspects that a fire has been deliberately set and requests assistance from the state fire marshal does so based on an examination of the scene and circumstances surrounding the incident. The chief's investigation may not have been an in-depth one, but without it, the further investigation would have been impossible. At the other end of the spectrum, many metropolitan fire departments have established dedicated fire investigation units. These units are typically part of the fire prevention bureau and are staffed with trained investigators who, in many cases, handle the investigation from the fire scene through the courts.

**NOTE** Fire departments participate in the investigation of fire to some extent in practically every jurisdiction.

The legal authority for fire departments to conduct fire investigations usually originates within a state statute. The chief of the fire department is either authorized or in some cases required to conduct or cause to be conducted an investigation into the cause and origin of every fire (Figure 9-1). The authority to establish an investigation unit with law enforcement powers also generally originates within statute, although in some instances, local investigation units actually derive its law enforcement authority from the state fire marshal. In these instances, local officials act as deputy state fire marshals.

**NOTE** The legal authority for fire departments to conduct fire investigations usually originates within a state statute.

**FIGURE 9-1**
The fire chief or incident commander typically conducts an initial investigation into the origin and cause of every fire.
(Courtesy of Duane Perry.)

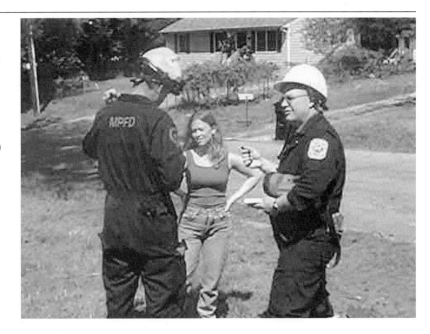

# Fire Prevention Bureau Investigation Units

The term *fire investigation* is a broad one and is composed of several functions. Some fire prevention bureaus perform some or all of these functions. Those that do not perform all of the functions rely on an outside agency, often the state fire marshal or state or local police department for assistance. Assistance from federal authorities, notably the Bureau of Alcohol, Tobacco, Firearms and Explosives (ATF), has become common in many jurisdictions, especially when dealing with incidents involving explosives, potential hate crimes, and/or federal crimes. U.S postal inspectors, the FBI, and other law enforcement agencies may also be involved.

The steely fire investigator portrayed in fiction is actually a disciplined, persistent seeker of the truth, willing to spend countless hours sifting through burned debris, performing countless interviews, and making an endless number of phone calls. The fire investigator plies their trade where it is wet, cold, dark, and far from glamorous (Figure 9-2). The heroes who extinguished the fire are back in a warm fire station, and the investigator is still at the fire scene, with water dripping from the ceiling and standing in puddles on the floors. The moon is shining through the hole in the roof, the stench of charred garbage permeates the building, and the cockroaches that peer from the shadows of the halogen lights are the only other living things left at the scene.

The scene is about as far from Hollywood as you can get. The job at hand is to determine where the fire started, who was involved, and why. The possibility of a conviction for arson can be lost if the fire scene is not thoroughly and effectively processed.

**FIGURE 9-2**
The fire investigator works in cold, wet, and dark conditions.
(Courtesy of Duane Perry.)

**NOTE** The possibility of a conviction for arson can be lost if the fire scene is not thoroughly and effectively processed.

NFPA's *NFPA 1033: Standard for Professional Qualifications for Fire Investigator* contains job performance requirements developed and maintained by NFPA's Technical Committee on Fire Investigator Professional Qualifications. The skills, which are considered nationally accepted good practice, cover the areas of scene examination, documentation, evidence collection, interview and interrogation, postscene investigation, and presentation.[5] *Fire Investigation* (Russ Chandler, Delmar Cengage Learning, 2008) is correlated with the performance requirements of *NFPA 1033*.

NFPA 921, *Guide for Fire and Explosion Investigations*, is a guide or how-to document for fire investigators. The preface states that the document is meant:

> ... to assist in improving the fire investigation process and the quality of information on fires resulting from the investigative process. The guide is intended for use by both public sector employees who have statutory responsibility for fire investigation and private sector persons conducting investigations for insurance companies or litigation purposes. The goal of the committee is to provide guidance to investigators that is based on accepted scientific principles or scientific research.[6]

## Certification

The International Association of Arson Investigators has maintained a national certification system for fire investigators since 1986 through its Certified Fire Investigator (CFI) and Fire Investigation Technician (FIT) programs. The CFI program is accredited by the

National Fire Board of Fire Service Professional Qualifications and provides for recertification through field work and continuing education. Certification involves a point system with credit for education, training and experience, and successful completion of a written test.

The National Association of Fire Investigators (NAFI) developed its certification program in 1982. Its certifications include Certified Fire and Explosion Investigator (CFEI), Certified Fire Investigation Instructor (CFII), and Certified Vehicle Fire Investigator (CVFI). The NAFI's certifications require attendance at a 4-day NAFI-sponsored course, a credential review, and successful completion of a written examination.

## THE INVESTIGATION PROCESS

The functions involved in the investigation of fires that can broadly be classed as investigation of the fire scene include interviews of witnesses; follow-up investigation, including interviews and examination of records and test results; securing and serving search and arrest warrants; and court preparation and testimony (Figure 9-3). It should be noted that none of these functions involves skills that the typical firefighter has developed in his or her career. All require initial and refresher training in order to be effectively accomplished. Additionally, most involve state-mandated law enforcement training and certification.

**NOTE** It should be noted that none of these functions involves skills that the typical firefighter has developed in his or her career.

**FIGURE 9-3**
Training for fire investigators includes courtroom testimony. (Courtesy of Duane Perry.)

# Investigation of the Fire Scene

The element of fire investigation most commonly performed by fire departments is the primary determination of the origin of the fire, or where it began, and the cause. In practically all fire departments, the initial officer in charge performs some introductory fire investigation. When the fire cause is not easily determined or appears suspicious or involves criminal activity, assistance is requested, and responsibility for the investigation is assumed by others with additional training and resources.

In some jurisdictions in which the fire department does not assume responsibility for the entire investigative process, specifically trained personnel are assigned to investigate and process fire scenes. When the cause is determined to be **incendiary**, the jurisdiction's law enforcement agency is called in to handle further investigation and secure search and arrest warrants and effect the arrest of those responsible. The fire department representative who processed the fire scene and determined the origin and cause of the fire becomes the **expert witness** for the prosecution, whose testimony will make or break the case. To be effective, a close working relationship between the fire and police departments is essential. The fire department personnel must be adequately trained and equipped.

**incendiary** a destructive fire that is intentionally set.

**expert witness** a person who by education or specialized experience possesses superior knowledge and is permitted to offer opinions during court testimony.

# Interviews and Follow-up Investigation

Effective interviews, evidence collection, and case management are skills often referred to as "old-fashioned police work." They are skills that must be learned and honed. The temperament and aptitude needed to develop into a first-class firefighter and those needed to do the job of a police investigator are not the same. Care should be exercised in selecting personnel to work as fire investigators. The word *fire* in front of the word *investigator* does nothing to lessen the potential hazards faced by a criminal investigator who just happens to be a member of the fire department.

**NOTE** The temperament and aptitude needed to develop into a first-class firefighter and those needed to do the job of a police investigator are not the same.

Some departments attempt to reduce the political impact that could be caused by the use of deadly force by a member of the fire department, but placing their personnel into situations in which they could easily encounter armed and dangerous felons without the benefit of arms for personal protection is unconscionable. Fire investigators should receive the same level of self-defense, firearms, and investigative and legal training as criminal investigators employed by police departments.

**FIGURE 9-4**
Training in self-defense is a crucial aspect of fire investigator preparation. (Courtesy of Duane Perry.)

**NOTE** Fire investigators should receive the same level of self-defense, firearms, and investigative and legal training as criminal investigators employed by police departments.

Interviewing witnesses, executing search and arrest warrants, giving effective courtroom testimony, using self-defense and firearms, and keeping up to date on legal issues require specialized training (Figure 9-4). State fire training academies, state fire marshal offices, the National Fire Academy, the Federal Law Enforcement Training Center, and the FBI Academy are just some of the organizations and institutions that provide effective training. Professional associations such as the International Association of Arson Investigators (IAAI), National Association of Fire Investigators (NAFI), and International Association of Bomb Technicians (IABTI) are among the groups that also provide training.

## Equipment and Resources

The types of equipment used by investigators on the fire scene are similar to those used at any crime scene, with additional items and equipment that reflect the hazardous nature of a fire scene and the fragility of the kinds of evidence that can be gleaned. Fire investigations texts such as Chandler's *Fire Investigation* and Kirk's *Fire Investigation* and IFSTA's *Fire Investigator Manual* provide useful lists of tools and equipment. Good working relationships with the jurisdiction's police department may enable the fire investigation unit to purchase supplies and equipment through the police agency, which can result in considerable savings and reduce the supplies needed for inventory. In addition to standard crime scene equipment, lighting equipment, evidence containers, shovels, screens for sifting debris, and other specialized items are needed for the fire scene. Fire scenes are photographed and sketched; fingerprints, tire, and footprints are sometimes recorded; and pieces of physical evidence are gathered, inventoried, and secured.

A secure location must be provided for storage to prove an unbroken **chain of custody** for all items. Contracts for the forensic testing of evidence and for processing photographs can often be established with police agencies.

> **chain of custody** system to verify by documentation that evidence has been in the sole control of law enforcement from the time it is seized to the time it is admitted in court.

The system established to investigate fires within many jurisdictions falls somewhere in between the fully staffed unit within the fire department and the unit completely outside the fire department. Many jurisdictions with fire prevention bureaus lack the resources to assign personnel to the full-time investigation of fires, either because of the frequency of fires or fiscal constraints. "Inspectigator" is a somewhat humorous term that is sometimes used to describe personnel charged with both fire inspection and fire investigation duties. Can an individual really excel at both? Yes, but to truly wear both hats, personnel must have the ability and aptitude to perform both jobs. They must also be trained and equipped for both.

The downside to this arrangement is that the skills needed for investigating and inspection are not the same. Personnel are forced to work all the harder at attaining and maintaining their edge. The upside is that the knowledge gained from each of the disciplines will improve performance in the other. An inspector who truly understands fire behavior and can use that knowledge to educate the business community while conducting inspections can accomplish a great deal. A fire investigator who has a good grasp of both building and fire codes can use that knowledge to better understand fire behavior and fire spread.

## Courtroom Testimony

Arson is unlike other crimes in that within the court system, even proving it occurred is often the most difficult task of the prosecution. To prove arson, all accidental causes must be effectively eliminated. Standard fire investigation reports contain information on the weather at the time of the fire. The notation addresses the possibility that a lightning strike was responsible for the fire. Similarly, electrical, mechanical, and other ignition possibilities must be eliminated as possible causes.

> **NOTE** To prove arson, all accidental causes must be effectively eliminated.

Defense attorneys must only establish the possibility that a fire was caused by some accidental means to move the jury to acquit their client. Imagine if that were the case with other crimes!

> DEFENSE ATTORNEY: Ladies and gentlemen of the jury, we will stipulate that my client was in the bank on October the 31st, and that he was armed with a handgun and was wearing a mask, and that when arrested he had $20,000, and a dye pack had exploded

covering his torso with red dye. The prosecution has failed to prove that my client is guilty. On the stand, the detective could not positively eliminate the possibility that my client wasn't on the way to a Halloween party, dressed as a bank robber, and that he stopped by the bank to get money for the collection plate for Mass this Sunday.

## Measuring Effectiveness

The measure of effectiveness for a fire investigation unit is the case closure rate, not the conviction rate. The best to be hoped for is that 100 percent of the fires be classified as accidental or incendiary, with every incendiary fire leading to an arrest. The rate of conviction lies with the prosecutor and is beyond the control of the unit.

**NOTE** The measure of effectiveness for a fire investigation unit is the case closure rate, not the conviction rate.

Arson is a complex crime that can be difficult to prove. A common tactic used by prosecutors is to offer to drop an arson charge in return for guilty pleas for other related crimes. If a criminal breaks into a residence, steals valuables, and sets a fire to cover the evidence of the crime, multiple offenses have occurred. If good investigative work at the fire scene leads to the arrest and conviction of the perpetrator, the case is a success, even if the arson charge is dropped in a plea agreement.

**NOTE** If good investigative work at the fire scene leads to the arrest and conviction of the perpetrator, the case is a success, even if the arson charge is dropped in a plea agreement.

## SUMMARY

The fire service has the greatest stake in determining the origin and cause of fires and channeling that information to prosecutors, government regulators, and the public. To effectively perform the function, personnel must be adequately trained and equipped and must work closely with law enforcement agencies and the court system. The measure of effectiveness is the case closure rate, not the conviction rate.

## REVIEW QUESTIONS

1. Which federal agency compiles the Unified Crime Reports submitted by jurisdictions within the United States?
2. What legal authorization is used by fire departments to form fire investigation units?

3. In addition to criminal prosecution, what are other uses for information from fire scene investigation?

4. List two federal law enforcement agencies involved in the investigation of fires.

5. List three professional associations that provide training for fire investigators.

6. List three federal training academies that provide fire investigator training for local officials.

7. Which two NFPA documents specifically address fire scene investigation?

8. What statistic is used to measure overall effectiveness of a fire investigation unit?

9. Why is conviction rate not a good indicator of a unit's effectiveness?

## DISCUSSION QUESTIONS

1. The fire chief is developing a proposal for the county council for the development of a fire investigation unit within the fire prevention bureau and is seeking the support of the firefighters' union. Develop a position paper with talking points for the chief to use in a presentation to union leaders. Describe how the unit will benefit the rank and file firefighter on the street.

2. The chief has informed you that he is leery of the potential political impact of armed fire investigators. Develop a list of points, both pro and con, for the chief to use in his deliberation.

## CHAPTER PROJECT

Using *NFPA 1033* as a guide, develop a list of basic skills needed by firefighters who are assigned to perform fire investigations and a list of skills needed by law enforcement officers who are assigned to perform fire investigations.

## ADDITIONAL RESOURCES

In-depth information on many of the subjects discussed in this chapter can be found in the following texts and publications and at these Web sites.

Russell K. Chandler, *Fire Investigation* (Clifton Park, NY: Delmar-Cengage Learning, 2009).

John DeHaan, *Kirk's Fire Investigation* (Upper Saddle River, NJ: Prentice Hall, 2006).

*Fire Investigator Manual*, Oklahoma City, OK, International Fire Service Training Association.

National Fire Protection Association at www.nfpa.org.

*NFPA 1033: Standard for Professional Qualifications for Fire Investigator* (Quincy, MA: National Fire Protection Association, 2009).

*NFPA 921: Guide for Fire and Explosion Investigations* (Quincy, MA: National Fire Protection Association, 2008).

U.S. Bureau of Alcohol Tobacco and Firearms at www.atf.gov.

U.S. Fire Administration at www.usfa.fema.gov.

# NOTES

1. James Leasor, *The Plague and the Fire* (New York: McGraw Hill, 1961), page 265.
2. Ibid., page 267.
3. *2008 Crime in the United States* (Washington, DC: Federal Bureau of Investigation, 2009), Table 27.
4. Ibid.
5. *NFPA 1033: Standard for Professional Qualifications for Fire Investigator* (Quincy, MA: National Fire Protection Association, 2009), sections 3-2 through 3-7.
6. *NFPA 921: Guide for Fire and Explosion Investigations* (Quincy, MA: National Fire Protection Association, 2008), page 1.

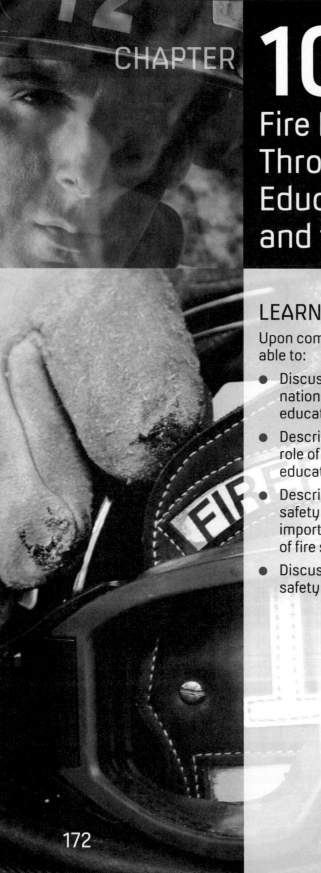

# 10

# Fire Prevention Through Public Education, Awareness, and the Public Forum

## LEARNING OBJECTIVES

Upon completion of this chapter, you should be able to:

- Discuss and contrast the use of local and national fire statistics in developing fire safety education programs.

- Describe the impact of *America Burning* on the role of fire departments in public fire safety education.

- Describe the five-step process for public fire safety education planning and discuss the importance of planning in the development of fire safety education programs.

- Discuss the role of the media in public fire safety education.

# PUBLIC FIRE SAFETY EDUCATION— SELLING FIRE SAFETY

For the most part, early fire prevention efforts in the United States were undertaken by the business community. Governments, particularly the federal government, got involved during times of war and national crises. Both business and government learned very quickly that to reduce the incidence of hostile fire, public support and cooperation were necessary. To garner public support, fire prevention had to be sold to the public. Ad campaigns, similar to those used to sell consumer items, were developed and used effectively to reduce the incidence of fire.

**NOTE** Both business and government learned very quickly that to reduce the incidence of hostile fire, public support and cooperation were necessary.

To many in the fire service, the term public fire safety education means Bert and Ernie and "Stop, Drop, and Roll" in elementary school classrooms. That certainly is a part of it, but the need for public fire safety education is much bigger and must be aimed at the entire public. Programs must be tailored to address specific problems and directed to those that have the power to affect change. Do not get carried away with the term *education*. *Sales* is probably the better term. Fire safety programs must appeal to the audience such that they identify with the program, see that it will improve their lives, and are willing to spend (in this case time or energy) to obtain the product (reduced fire loss and personal safety). Just as Madison Avenue appeals to the public's emotions, fire safety education programs must have a hook embedded within a bait that gets the public's attention and makes them bite.

Think back to an effective advertising message that got your attention. The antidrug message that showed two eggs frying in a skillet with the voice-over "This is your brain on drugs" got most people's attention. It said a lot with very few words. It appealed to your emotions. It made you not want to do something. A classic fire prevention message that had great success in the 1960s, was the National Board of Fire Underwriters' (NBFU) "Stop the Fifth Horseman," a full-page ad that appeared in national magazines during Fire Prevention Week (Figure 10-1). With the Four Horsemen of the Apocalypse in the background, Fire, personified as the fifth horseman astride a black horse, waved his torch of destruction. The biblical reference to the Four Horsemen struck a chord with most of the public. The ad probably would not have the same effect today as it once did. A lot of young kids would probably think of the fifth horseman as a "righteous dude."

Patriotism, fear, and the desire for economic prosperity have all been used with varying degrees of success. Many factors influence the potential success of the effort. Age and a multitude of socioeconomic factors have a huge impact. Misreading or not understanding the audience can be disastrous. The National Fire Protection Association (NFPA)'s Learn Not To Burn™ curriculum, designed for use in schools for preschoolers through eighth grade, is billed as positive and nonthreatening. Dr. Frank Field, a long-time TV meteorologist and science reporter in New York, believed that in intending to be nonthreatening, most fire safety education programs were too simplistic and were not up to the job.

**FIGURE 10-1**

"Stop the Fifth Horseman" was a successful fire prevention campaign of the National Board of Fire Underwriters. (Courtesy of the American Insurance Association, Washington, D.C.)

When he reviewed the fire safety education materials that his grandchildren were receiving, he decided to take on the problem.

He had some background. In the late 1980s as science editor at a WCBS TV in New York, Dr. Field investigated disturbing trends in the fire safety records of major U.S. cities. He found that in terms of fire safety, it was safer to live in a foreign city such as Tokyo, than in a major U.S. metropolitan area. He partnered with his son, Storm Field, and daughter, Alison Field, who are also broadcast journalists, and with a grant from the MetLife Foundation, developed a DVD titled *FIRE IS*. The program is made up of four 20-minute segments titled "Fire is Black," "Fire is Hot," "Fire is Fast," and "Fire is Smoke & Gas and Fire is an Emergency." The DVDs are available in both English and Spanish.

After a screening for firefighters, Steve Cassidy, president of the 24,000-member Uniformed Firefighters Association of New York, said:

> New York City Firefighters risk our lives every day for the safety of others and this DVD can go a long way toward preventing unnecessary injury and death to civilians and firefighters alike. This is far and away the best fire safety education tool available today.

The UFA underwrote the cost and distributed 2,000 copies to New York City Schools. *FIRE IS* is realistic, but the developers were careful to keep the program nonthreatening.

Preying on children's fears not only reduces the likelihood of success with the children; it also sets the stage for real problems with parents. It only takes a few calls to the principal from angry parents before fire safety education is no longer a welcome topic.

**NOTE** It only takes a few calls to the principal from angry parents before fire safety education is no longer a welcome topic.

Tailoring classes and fire safety messages to specific high-risk target groups has proven effective. Senior citizens have statistically higher rates of death and injury from fire. If you are fortunate enough to celebrate your eighty-fifth birthday, your odds of dying by fire will increase by nearly fivefold over that of a 20-year-old person.[1]

**NOTE** If you are fortunate enough to celebrate your eighty-fifth birthday, your odds of dying by fire will increase by nearly fivefold over that of a 20-year-old person.

> When we come to the question of fire prevention, there are three main points to consider. First, a good citizen will try in every way to avoid being a cause of danger through permitting any of the practices that we have been warned about. Second, he will remove all dangerous conditions that he may find in his home, and third, he will train himself to recognize dangerous conditions in the community and will use his influence both to have them removed and to educate others to the habits of carefulness.[2]
>
> From Safeguarding the Home Against Fire, A Fire Prevention Manual for the School Children of America, 1918.

## EDUCATION, THE SECOND "E"

"Engineering, education, and enforcement," the three E's of fire prevention, is an old slogan that continues to be used throughout the fire service (Figure 10-2). Educating business owners and the public continues to be one of the most effective fire prevention strategies. The NFPA was formed in 1896 by engineers representing stock insurance companies in New England. Many insurance companies had developed their own standards for the installation of automatic sprinklers, and there was a problem. Nine radically different standards were in use within 100 miles of Boston.[3] The association was formed to "Promote the science and improve the methods of fire protection, to obtain and circulate information on this subject, and to secure cooperation in matters of common interest."[4]

The NFPA quickly began the formulation of engineering standards, beginning with automatic sprinklers. In short order, standards for the protection of openings in walls and partitions, fire pumps, signaling systems, hose and hydrants, and other fire protection features were developed. In 1900, the NBFU joined the NFPA and voted to adopt the NFPA's standards and assume the cost of publishing them.[5] The NFPA standards were originally published as NBFU pamphlets and carried the subtitle "as recommended by the National Fire Protection Association."

**FIGURE 10-2**

Education is the second of the Three E's. (Courtesy of New York University.)

*In Fifty Years of Civilizing Force,* the early history of the NBFU, Harry Chase Brearley describes how NFPA was thrust into the business of fire safety education:

> The NFPA was full of energy and devotion, and, in its somewhat academic way, it felt well satisfied with its efforts toward the reduction of fire-waste until about five or six years ago [approximately 1910], when there came a rude awakening. Someone arose in one of its meetings and called attention to the fact that the fire-waste was not being reduced—that, on the contrary, it was increasing rapidly. This caused anxious thought. The association knew that it had worked out the physical standards which would curtail this waste—if applied. Failure to decrease the waste indicated failure to apply the standards on the part of the public. The people of the nation had not yet been aroused to the urgency of the situation. Someone must assume the important and most obvious task of arousing them. Who? Obviously, the body which had studied the extent of the fire-waste, which had worked out the standards, and which knew how easily they might be applied. In other words, the National Fire Protection Association had had a burden laid upon it in its very name; it now saw the necessity for adding a course in human engineering to its already established work in mechanical and structural engineering.[6]

The NFPA began its program of "human engineering" in 1910 by distributing press bulletins to newspapers throughout the United States. Its first attempt was almost a total failure; only a single U.S. newspaper, the *Boston Herald*, agreed to publish its fire safety bulletins. NFPA Secretary Frank Wentworth quickly realized that to succeed, the NFPA would have to sell the idea of fire prevention, and he enlisted the cooperation of an organization that had the ear of almost every newspaper in the country, the National Association of Credit Men.[7] Harry Chase Brearley identified the great lesson learned by the NFPA in its partnership with the National Association of Credit Men, described as the "first great lesson of publicity"—translate the abstract into concrete.[8]

The lesson is as valid today as it was almost a century ago. For a fire safety message to be effective, members of the public must personally identify with it. Reducing the fire threat within the community to ensure economic prosperity is an abstract. Most would agree it must be a good idea, but it is not much different from wishing for good weather. But when the 6 o'clock news contains a segment with footage from an industrial fire and then interviews with factory workers who are subsequently unemployed and applying for public assistance, people pay attention. Translating the abstract into concrete means putting faces on the victims, making them real people. It also means striking while the iron is hot—the best time to speak about the importance of smoke alarm maintenance is while the debris from the fire is still smoking in the front yard.

**NOTE** For a fire safety message to be effective, members of the public must personally identify with it.

**NOTE** The best time to speak about the importance of smoke detector maintenance is while the debris from the fire is still smoking in the front yard.

## National Fire Prevention Programs

Fire Prevention Day was first observed on October 9, 1911, the fortieth anniversary of the Great Chicago Fire, at the suggestion of the Fire Marshal's Association of North America.[9] The NBFU approached state governors, and many issued Fire Prevention Day proclamations.[10] President Woodrow Wilson issued the first National Fire Prevention Day Proclamation in 1920. President Warren Harding officially proclaimed the first Fire Prevention Week in 1922 with the statement: "Fire Prevention Week is to be observed by every man, woman, and child, not only during the week designated in this pronouncement but throughout every hour of every day of every year."[11]

In 1918, the NBFU produced and distributed more than 2 million copies of a 91-page booklet titled *Safeguarding the Home Against Fire*, through the United States Bureau of Education (Figure 10-3). Developed in cooperation with the National Association of Credit Men, the pamphlet was distributed to elementary schools throughout the country.[12]

The pamphlet included fire-loss comparisons between the United States and Europe included in Figure 10-4.[13] It then translated the abstract into concrete with the following

**FIGURE 10-3**

Over two million copies of *Safeguarding the Home
Against Fire* were distributed to schools in 1918.

(Source: *Safeguarding the Home Against Fire*, National
Board of Fire Underwriters, 1918.)

**FIGURE 10-4**

Per capita fire loss in 1913 in seven countries.

| NATION | FIRE LOSS IN 1913 |
|---|---|
| Holland | $ 0.11 |
| Switzerland | $ 0.15 |
| Italy and Austria | $ 0.25 |
| Germany | $ 0.28 |
| England | $ 0.33 |
| France | $ 0.49 |
| United States | $ 2.10 |

quotation from a 1909 address titled "Conservation of Natural Resources" by Charles
Whiting Baker, editor of *Engineering News*:

> The buildings consumed [by fire in 1909] if placed on lots of 65 feet frontage, would
> line both sides of a street extending from New York to Chicago. A person journeying
> along this street of desolation would pass in every thousand feet, a ruin from which an
> injured person was taken. At every three-quarters of a mile in this journey he would
> encounter the charred remains of a human being who had been burned to death.[14]

Twelve years later, *Safeguarding the Nation Against Fire*, a 132-page pamphlet directed at
high school students, was produced and distributed by the NBFU (Figure 10-5). Again,
the cost was borne solely by the insurance industry without government assistance.
Altruism? Yes, but good business in the truest sense.

**FIGURE 10-5**

Over two million copies of *Safeguarding the Nation Against Fire* were distributed in 1930. (Source: *Safeguarding the Nation Against Fire*, National Board of Fire Underwriters, 1930.)

The stiff-collared insurance executives believed in God, country, and the system of free enterprise without the clear-cut divisions between the three that have become established in our current society. Being a good citizen and working hard to subdue the land and make it fruitful were all part of the big plan. Without credit to purchase land and equipment, industry and commerce were impossible. Without insurance to protect the property that was the basis for loans, credit was not possible. In the end, commerce, good citizenship and, of course, fire prevention were all part of God's plan for America.

## PLANNING FIRE SAFETY EDUCATION PROGRAMS

To be effective, fire safety education programs must target audiences that can have an impact on the jurisdiction's fire problem. Every community has high-risk areas, groups, or conditions that warrant the efforts of fire prevention bureaus. National statistics gleaned from the U.S. Fire Administration (USFA) and NFPA reports identify trends and conditions within the United States, often broken down by state or region.

> **NOTE** To be effective, fire safety education programs must target audiences that can have an impact on the jurisdiction's fire problem.

Developing fire safety education programs based solely on the fire experience of the jurisdiction suffers from the same flaw recognized in Chapter 6 regarding fire safety inspections. The very size of the statistical sample within a single jurisdiction is so small that it may not reflect an accurate picture of the potential hazards. Across the United States, senior citizens older than 70 years of age account for about 26 percent of the fire fatalities each year. Children younger than 5 years of age account for about 7 percent. Together, that reflects about one-third of all fire fatalities each year.[15] The fact that there have been few fatal fires in your jurisdiction involving senior citizens or young children probably does not mean that they are not the highest risk groups in your community.

Your local statistics should be carefully reviewed because you may well have specific conditions that result in higher risks than those of the state or region. The Atlanta Fire Department probably does not need to conduct a campaign to convince the public to clear the brush away from their homes, but that is a top priority for Southern California fire departments. Fires involving portable heating equipment might cause the Atlanta Fire Department to develop a program to educate the public in the proper use of space heaters, a program that might not be needed in San Diego. Both jurisdictions probably have a similar fire experience involving senior citizens who live alone or in assisted living facilities being at higher risk from fire.

> **NOTE** Your local statistics should be carefully reviewed because you may well have specific conditions that result in higher risks than those of the state or region.

In the National Fire Academy course Fire Prevention Organization and Management, two ingredients are identified as essential to effective planning for public fire safety education: planning and people.[16] Effective planning is crucial to the success of any program. People, those who are selected to develop and deliver the message, as well as key members of the public, the business community, civic and social groups, and other government organizations are the backbone of the program. Without their support for implementation, the program will fail. The selection of fire service personnel for fire safety education duties is as important as the message. Sample job descriptions for fire safety educator and public information officer (PIO) are included in NFPA standard *NFPA 1035: Standard for Professional Qualifications for Public Fire and Life Safety Educator.* As in other NFPA standards for professional certifications for fire service personnel, key knowledge, skills, and abilities (KSAs) can be found within the standard's job performance requirements (JPRs).

# The Five-Step Program

*Public Fire Education Planning, A Five-Step Process* was developed and released by the U.S Fire Administration in 1979 as a guide for planning public fire safety education programs. A step-by-step process designed to walk planners through each phase of a fire safety education program, the guide is available free of charge from the USFA's Publications Center. The five steps are identification, selection, design, implementation, and evaluation (Figure 10-6). If you picture the steps in a line, you are assuming that the five-step process is linear and has fixed start and end points. However, the process is actually a loop, with evaluation as the fifth step that leads back to step one, identification.

## Identification

Identification is the process of using fire records and statistics to determine the most serious fire problems facing the community. As previously discussed, high-risk groups and locations are not necessarily the ones in which your community has experienced losses in the recent past. A thorough review of local, regional, and national statistics should be undertaken. Reports from the USFA, the NFPA, the state fire marshal, and your jurisdiction should be studied.

Good regional cooperation among department and state agencies provides enormous benefits in all facets of fire prevention. There may already be a regional body formed to cooperate in the areas of fire investigations or inspections. If the bridge is already there,

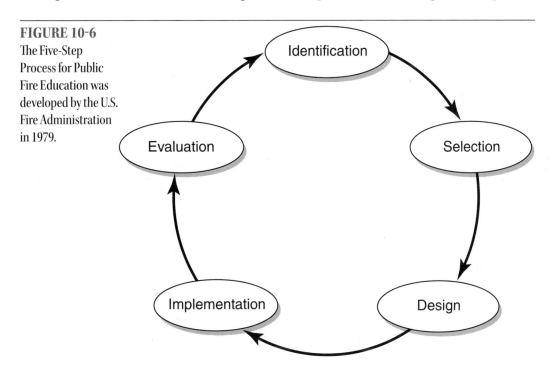

**FIGURE 10-6**
The Five-Step Process for Public Fire Education was developed by the U.S. Fire Administration in 1979.

do not build a new one. Rather, use the contacts within the organization to establish working relations among personnel conducting the public education programs.

## Selection

After the major high-risk persons or occupancies have been identified, the target audiences must be selected (Figure 10-7). An important point to remember is that high-risk persons may not necessarily be the target audience. The target audience is the person or persons who can take whatever actions are necessary to effect the desired change. Programs geared to address safety issues that confront young children and many seniors may target their caregivers. Programs that seek to reduce injuries from fireworks may target elected officials in an effort to reduce the availability of fireworks through legislation.

A key element in the selection process is the identification of available resources, both material and human. Assistance in the form of funding; staffing; and other essentials such as free advertising, printing, or videographic assistance should be identified and quantified. The key of which target audience can give you the most bang for your buck must be a driving force in the selection process.

## Design

In this step, an actual program is developed for presentation to the target audience. Whether the program is a series of commercials designed to be played by local radio stations, a specific presentation designed to be presented before a live audience, or

**FIGURE 10-7**
Cooking fires are a leading cause of property damage and can be reduced through educational programs.

a program designed to be handed off and presented by another group or groups, the message is packaged into a the delivery system for the target audience that was selected in step 2.

## Implementation

During the selection step (step 2), one of the resources that was identified and selected was the method of delivery for the program. Taped public service announcements prepared for airing by local radio stations are of little value if the station management is not on board before the ads are written. Accurately defining the delivery method is just as important as identifying the target audience. The answers to who, where, and how, must be in place before the design step (step 3) because programs are written for delivery by specific groups and for delivery to specific audiences.

During the implementation step, the program may have to be adjusted or tweaked to adjust for conditions that were not anticipated or expected. During the first few deliveries, instant feedback may reveal tremendous opportunities that were not anticipated. Conversely, problems may be identified that can be quickly addressed. Public fire safety education is a lot like sales and a lot like show business. Public fire safety educators must be able to think on their feet and make adjustments to ensure the message is delivered, received, and accepted.

**NOTE** Public fire safety educators must be able to think on their feet and make adjustments to ensure the message is delivered, received, and accepted.

## Evaluation

Evaluation is the fifth, but not the final, step. As previously mentioned, the process is a loop or a cycle. Without some method of evaluation, the effectiveness of programs cannot be measured. Evaluation provides a view of how the program affected changes in awareness, knowledge, or behavior. It provides a mechanism for program managers to make changes that will increase the effectiveness of the program.

The ultimate measure of success would be a drastic decrease in the number of fires or number of injuries or deaths. Unfortunately, the same limitations that made your local statistics too small a sample to use as the sole basis for determining your target audience apply to their use as the sole indicator of your success. You can do everything right and still not have the results reflected by a significant decrease in your local fire statistics. The statistical sample is just too small; other methods must be used to measure the effectiveness of your programs. Measuring changes in the level of knowledge or awareness of the target audience is a legitimate indicator as are changes in behaviors.

One method of determining effectiveness is through the use of tests or surveys before and after the program. These may be administered to the audience at group presentations or to representative samples of the general population after a media campaign. If certain behaviors are known to contribute to the fire problem and evaluation indicates a reduction in those behaviors, the program has been successful.

# FIRE SAFETY EDUCATION IN THE SCHOOLS

Fire safety education within the schools was recognized almost a century ago as an effective method of instilling fire-safe behaviors and attitudes in our children. The NBFU's *Safeguarding the Home Against Fire* and *Safeguarding the Nation Against Fire* were two early campaigns that were aimed at the children of the United States through the cooperative efforts of business and the U.S. Bureau of Education.

In 1942 the NBFU, together with the NFPA, Boy Scouts of America, Girl Scouts of America, National Fire Waste Council, and U.S. Department of Agriculture, developed and distributed a program through the New York University Center for Safety Education. The program, titled *Fire Prevention Education*, was packaged in a 350-page text and included a program for schools and the community with available resources listed in appendices. What is truly astounding is that in 1942, 70 fire safety education films were available for loan, free of charge, from the NBFU, NFPA, and Department of Agriculture.[17]

Fire prevention education as an important part of a fire department's mission is often traced back to the 1973 report of the National Commission on Fire Prevention and Control.[18] The report, titled *America Burning*, contained specific recommendations regarding fire safety education within schools:

- The Commission recommends that the Department of Health, Education and Welfare include in accreditation standards fire safety education in the schools throughout the school year. Only schools presenting an effective fire safety education program should be eligible for any Federal financial assistance.[19]

- The Commission recommends to the States the inclusion of fire safety education in programs educating future teachers and the requirement of knowledge of fire safety as a prerequisite for teaching certification.[20]

Public education by fire departments existed long before the 1973 *America Burning* report, as evidenced in the 1942 program, *Fire Prevention Education*, although not to the extent we see today. The program included a section titled "Classified List of Selected Publications on Municipal Fire Prevention and Fire Protection," which contained films, handouts, demonstrations, plays, games, and contests for use by municipal fire departments for public education.

Programs continue to be developed for use within schools (Figure 10-8). The NFPA's *Learn Not To Burn*® was first released in 1979. The program teaches 22 key fire safety features and is organized into three learning levels. The program is intended for use by individual classroom teachers and incorporates fire safety behaviors into regular school subjects, effectively incorporating fire safety education into the entire school year. *Risk Watch*® is an injury prevention program designed for use in classrooms for prekindergarten through eighth grade. The NFPA developed the program to address the large number of deaths and injuries that result from natural disasters and accidents involving firearms, poisoning, scalds and burns, falls, drowning, and others.

**FIGURE 10-8**
Miss Debbie discusses fire safety with Hector the Smoke Detector during a school program. (Photo courtesy of Scott Boatright.)

## COMMERCIAL FIRE SAFETY PROGRAMS

Few organizations have the staff resources to develop and maintain complex fire safety programs. Most organizations juggle staff and assignments to free personnel for fire safety education. Most chiefs recognize the benefits of fire safety education as an effective method of fire prevention and as good advertising for the department. Programs designed for delivery within school systems must meet rigid educational standards. In most cases, the most effective method of providing complex programs is by using those that are commercially developed.

**NOTE** In most cases, the most effective method of providing complex programs is by using those that are commercially developed.

The USFA develops hand-off educational programs geared toward fire prevention and safety. Public education pamphlets and materials can be obtained through the USFA Publications Center or ordered online. The Fire Safety Directory is a list of materials and resources available from other organizations ranging from burn and scald prevention to electrical hazards. The USFA maintains the list to assist agencies interested in developing public education programs.

NFPA's *Learn Not to Burn*® and *Risk Watch*® programs are aimed at fire and injury prevention for children and are widely used throughout the United States. *Remembering When: A Fire and Full Prevention Program for Older Adults*™ is a program aimed at reducing injuries and deaths in senior citizens from falls and fires. The downside is that these

**FIGURE 10-9**
High-quality fire educational materials are commercially available.

products are not cheap. One of the biggest challenges in providing fire safety education is the cost of the materials. The NFPA's *RISK WATCH* program materials cost $115 for each class of 30. FIRE IS is available to view free online or can be purchased for a small handling fee (Figure 10-9).

Fire safety education products are developed by commercial entities, either for sale or under government contract for free distribution. Fireproof Children Company provides programs in general fire safety and juvenile firesetter intervention and prevention. *Play Safe! Be Safe!* is an educational program for children ages 3 to 5 years for use in preschools and daycare centers and by firefighters at presentations.

## FIRE SAFETY EDUCATION FOR ADULTS

Fire safety education for adults is often accomplished through the media or by using conduits such as their children to deliver messages. "Change your clock, change your smoke detector battery" is heard in local newscasts across the country twice a year. Skeptics tend to dismiss the effectiveness of the message, but it does produce results, especially among those who are conscientious and simply never think about their smoke alarms. The message is enhanced when children, who received a fire safety lesson in school that week, ask their parents about the smoke alarms in their home.

Fire Prevention Week messages and press releases are prepared by the NFPA and the USFA, and many news organizations approach local fire departments for news spots concerning fire prevention topics. For the most part, adults are not the captive audience found in schoolrooms, but opportunities are available to educate adults, and many of them are not exploited to their full potential.

## Fire Safety Education for Businesses

Fire codes mandate employee and occupant training and the development of emergency plans for certain businesses. The examples of the types of buildings and occupancies in Figure 10-10 were taken from the International Fire Code.[21] The codes require the development of fire safety and evacuation plans and mandate employee training and drills in which the plans are exercised or tested. There is no better opportunity for fire safety education of adults and for establishing good relationships with the business community. Rather than simply demanding the development of a fire safety plan, fire prevention bureaus should develop boilerplate plans that can simply be filled in by business owners or representatives.

**NOTE**    Rather than simply demanding the development of a fire safety plan, fire prevention bureaus should develop boilerplate plans that can simply be filled in by business owners or representatives.

The Fire Department City of New York (FDNY)'s Bureau of Fire Prevention distributes a 14-page fire safety plan for high-rise hotels and a plan for high-rise office buildings. The plans can be downloaded as PDF files and filled in by hotel safety managers. The beauty of boilerplate plans is that they take the guesswork out of preparation, which saves business owners and the fire prevention bureau valuable time.

## Motivating Adults

Fire safety planning, training, and drills that involve building occupants are some of the most effective methods of improving the chances for appropriate response from building occupants in case of a fire emergency. Unfortunately, even the best efforts do not ensure whole-hearted participation from building occupants. Sometimes it takes a real fire to bring the very real possibility of an emergency into clear focus. Perhaps the best example was a recent one—the bombing of the World Trade Center (WTC) in 1993, followed by the terrorist attacks on September 11, 2001.

The WTC, a seven-building complex owned and operated by the Port Authority of New York and New Jersey, instituted a fire safety plan that included fire wardens on each floor, training for all employees, and semiannual drills. Because it was owned and operated by a government entity, the WTC was exempt from New York City's building and fire codes. In the postbomb report prepared by the WTC Risk Management Staff, which was published in *Fire Engineering Magazine* and later became USFA *Technical Report 076*, the staff described the fire safety plan:

> Each floor had fire safety wardens, provided with specific training to understand their duties during drills and actual emergencies . . . WTC fire drills are more complex than those required by the City of New York.[22]

**FIGURE 10-10**

Training and drills are required for these occupancies by the model codes. (Based on the International Fire Code.)

| OCCUPANCY/ BUILDING | TYPE | EXAMPLES |
|---|---|---|
| Group A | Assembly occupancies | Theaters, television studios, banquet halls, nightclubs, restaurants, bars and taverns, auditoriums, bowling alleys, courtrooms, dance halls, funeral parlors, gyms, libraries, museums, passenger terminals, arenas, houses of worship with over 2,000 occupants |
| Group E | Educational occupancies | Schools (K–12th grade), day-care facilities for more than five children, over $2\frac{1}{2}$ years |
| Group H | High-hazard occupancies | Buildings and structures containing materials in excess of permissible amounts, listed in fire and building codes, and semiconductor manufacturing and research |
| Group I | Institutional occupancies | Residential board and care, assisted living, half-way houses, group homes, alcohol and drug treatment centers, convalescent homes, hospitals, nursing homes, sanitariums, mental hospitals and detoxification facilities, child care facilities for children under $2\frac{1}{2}$ years, prisons, jails, corrections facilities |
| Group R-1 | Transient multifamily residential occupancies | Hotels, motels, boarding houses |
| Group R-4 | Assisted living | Assisted living facilities with over 16 residents |
| High-rise buildings | All used groups | Buildings with occupied floors more than 75 feet above the lowest level of fire department vehicle access |
| Group M | Retail stores, shops where stocks of goods are displayed and sold | Stores and shops with occupant loads greater than 500, or than 100 persons above or below the level of exit discharge |
| Covered malls | Buildings that enclose a number of tenants such as retail stores, drinking and dining establishments, entertainment and amusement facilities, and passenger terminals | Covered malls exceeding 50,000 square feet |
| Underground | All uses | Buildings with a floor level used for human occupancy more buildings than 30 feet below the lowest level of exit discharge |
| Buildings with atriums in Groups A, E, or M | Assembly education or mercantile buildings with atriums | An opening connecting two or more floors other than enclosed stairs, elevators, or utility shafts |

**FIGURE 10-11**
The World Trade Center bombing in 1993 exposed flaws in the complex's fire safety and evacuation plan.

On February 26, 1993, a truck bomb exploded in an underground parking garage in the WTC (Figure 10-11). The explosion and ensuing fire killed six and injured more than 1,000. In the aftermath, Anthony Fusco, chief of the FDNY, called the incident "the largest incident ever handled in the department's 128 year history."[23] Fifty thousand people were evacuated, making it the largest single evacuation ever recorded at the time. Search and evacuation were completed 11 hours after the explosion.[24]

Despite a complex plan with fire wardens and drills twice annually, the evacuation was neither smooth nor rapid. Many of the floor wardens were unfamiliar with stair locations; many had never left their floors during drills. In a study of the evacuation by the NFPA, some of the fire wardens reported that they were simply trained to "meet in the corridor and await instructions."[25] It becomes obvious that most of the employees never actually thought they would have a fire, much less be the victims of a terrorist attack.

On September 11, 2001, the unthinkable happened. Fully loaded jumbo jets were flown into the towers, resulting in tremendous fires and the eventual collapse of both 110-story towers. Of the 58,000 persons estimated to be at the WTC Complex, 2,830 lost their lives, including 403 emergency responders.[26] The North Tower was struck at 8:45 AM and collapsed in 103 minutes. The South Tower was struck at 9:03 AM and collapsed in 62 minutes.

Figure 10-12 lists the evacuation times reported by WTC occupants after their evacuation at the 1993 bombing.[27] Eight years later, on September 11, 2001, evacuation times were obviously very different. The towers may not have been fully occupied because of the hour of the morning, but that is not the point. Consider what could have happened had occupants of the complex not promptly evacuated. Compare the collapse times for the towers: 103 minutes for the North Tower and 62 minutes for the South Tower.

**FIGURE 10-12**

Evacuation times after the World Trade Center bombing tragedies in 1993.

| TIME REQUIRED TO EVACUATE | WTC NORTH TOWER | WTC SOUTH TOWER |
|---|---|---|
| Less than 5 minutes | 1% of the occupants | 1% of the occupants |
| 5 to 30 minutes | 13% of the occupants | 23% of the occupants |
| 30 to 60 minutes | 26% of the occupants | 47% of the occupants |
| 90 to 180 minutes | 52% of the occupants | 28% of the occupants |
| Over 180 minutes | 9% of the occupants | 1% of the occupants |

Instead of the 2,830 people who lost their lives, what if it had been 35 percent of the 25,000 people in the North Tower and 29 percent of the 25,000 people in the South Tower? The incident was the most horrific thing to befall the nation and the fire service. How much worse would 16,000 fatalities have been? It is obvious from the number of fatalities on September 11, 2001, that building occupants evacuated much more quickly than during the 1993 bombing. Did the reality of the bombing raise the level of awareness in the WTC and cause occupants to take the WTC fire safety plan and drills more seriously? The numbers say yes.

**NOTE** It is obvious from the number of fatalities on September 11, 2001, that building occupants evacuated much more quickly than during the 1993 bombing.

**NOTE** Did the reality of the bombing raise the level of awareness in the WTC and cause occupants to take the WTC fire safety plan and drills more seriously? The numbers say yes.

In the wake of fatal fires and explosions, some would say that using the incident as a method of bringing home the reality of fire brings unnecessary pain to the survivors and family members of the victims. To the contrary, fire safety education on the heels of fatal incidents is one of the few ways to bring good out of bad situations and should be considered a tribute to the victims.

**NOTE** Fire safety education on the heels of fatal incidents is one of the few ways to bring good out of bad situations and should be considered a tribute to the victims.

## EFFECTIVE USE OF THE MEDIA

The media has been both reviled and idolized at various times in history, probably more often the former. In addition to its mission of reporting the news to the public, the media also provides significant amounts of information to the public regarding everything from quality of life issues to safety. Using the media as a conduit for fire safety education in both news reporting and information dissemination has the potential for reaching the largest audience that extends across all levels of society.

**NOTE** Using the media as a conduit for fire safety education in both news reporting and information dissemination has the potential for reaching the largest audience that extends across all levels of society.

Public officials and agencies that refuse to cooperate with the media by stone-walling or simply refusing to answer calls often get a rude awakening. When accurate information cannot be obtained through normal channels, the media resorts to gleaning information from unofficial sources, often resulting in less-than-accurate information reaching the public. There is much to be gained by establishing good working relations with the media, not the least of which is limiting the amount of inaccurate information reported about the fire department.

**NOTE** When accurate information cannot be obtained through normal channels, the media resorts to gleaning information from unofficial sources, often resulting in less-than-accurate information reaching the public.

Some fire departments have the luxury of sufficient staffing to assign personnel to regular duty as PIO, which enables the media and the department to forge working relationships. *Forge* is a better word than *establish* in this case because the working relations are sometimes steeped in conflict arising from the media's need to push to get the facts. That conflict is a fact of life and is to be expected. PIOs who can comfortably work within that environment can establish media ties based on mutual respect and honesty.

There is no better platform than the fire scene to deliver the message of fire safety. However, care must be taken to ensure that the release of information and access to fire scenes does not compromise the fire investigation process or pose a safety hazard. If the fire prevention bureau is not the public information contact for the fire department, then its contacts with the press should be coordinated with the department's PIO. When poor coordination exists between branches, the stage is set for the press to shop around for the right answer.

## PUBLIC HEARINGS

Public hearings are a fact of life for public officials. Fire chiefs must justify their budgets during public hearings before elected officials. Fire prevention bureaus become involved during hearings on code adoptions and at hearings before appeals boards. Local cable access channels have added a new twist to the process—these hearings are often televised, opening the process to a wider audience.

## Hearing Preparation

The best public officials are those who can think on their feet and are totally prepared. The more time spent in preparation for public hearings, the less there is to go wrong. Fire prevention bureau members who are charged with the preparation of hearing documents and scheduling must use great care in document preparation and follow the legal requirements regarding advertisement of the hearing.

**NOTE** The more time spent in preparation for public hearings, the less there is to go wrong.

State and federal Freedom of Information and Sunshine laws prohibit the deliberation of elected officials except in duly advertised public hearings. Fire department managers should have a good working knowledge of the regulations. These laws also apply to documents in the possession of fire prevention bureaus. Chapter 11, "Fire Prevention Records and Record Keeping," discusses these laws in greater detail.

## SUMMARY

Public fire safety education is a lot more about sales than education. Education has been termed the "central factor" in effective fire prevention programs and is always listed as the second, or middle, element in the three E's of fire prevention, which consist of engineering, education, and enforcement. Fire protection and prevention specialists have long realized that engineering solutions were ineffective without the cooperation of the public. To gain that cooperation, carefully targeted fire safety education programs aimed at those who can effect necessary changes must be delivered.

Public fire safety education is not a modern phenomenon. Programs were instituted in the early 1900s by the insurance industry through the NBFU and the NFPA. State governments were the first to recognize Fire Prevention Day in 1911, which was the brainchild of the Fire Marshal's Association of North America, and used the fortieth anniversary of the Great Chicago Fire as a springboard. The event was first recognized by the federal government in 1920 and was expanded to Fire Prevention Week in 1922.

Although fire departments have been involved in fire safety education for more than a century, the 1973 report of the National Commission on Fire Prevention and Control, titled *America Burning*, is recognized as a watershed for public fire safety education by fire departments. The report recognized the need for fire safety education and recommended the creation of the USFA and National Fire Academy. USFA fire safety education programs, which are available for use by fire departments free of charge, and the public fire safety education curriculum at the NFPA can be traced back to *America Burning*.

Effective use of the media is a critical element in the delivery of effective fire safety education programs. Effective working relationships must be forged and maintained because no other method of delivery has the potential to reach as broad a segment of the population.

## REVIEW QUESTIONS

1. Why is it inadvisable to use local fire statistics as the sole method of identifying high-risk areas, groups, or conditions?
2. What was the title of the 1973 report of the National Commission of Fire Prevention and Control?
3. List the steps identified in the five-step process for public fire safety education.
4. What federal agency was created in response to the 1973 report of the National Commission of Fire Prevention and Control?
5. List two functions performed by the agency.

## DISCUSSION QUESTIONS

1. A local service club has donated several thousand dollars to the fire department to be used for a fire safety education program that benefits the local community. The chief has directed you to prepare a presentation to be given at the club's next dinner meeting outlining the method used to determine the target audience. Briefly describe strategies to determine the audience.
2. You are directed to attend a meeting at the mayor's office to discuss the development and implementation of a program to reduce the incidence of fires and burn injuries involving residents with Alzheimer's disease in local nursing and care facilities. The mayor wants the department to develop a series of classes to be delivered to these residents within the care facilities and is especially interested because her mother has the condition. Is this the best method of reducing fires and injuries involving this high-risk group? What are your recommendations to the mayor?

## CHAPTER PROJECT

Using *NFPA 1035* as a guide, develop a list of basic skills needed by firefighters who are assigned to perform public education programs.

## ADDITIONAL RESOURCES

In-depth information on many of the subjects discussed in this chapter can be found in the following texts and publications and at these Web sites.

*FIRE IS* at www.fireis.com.

National Fire Protection Association at www.nfpa.org.

*Fire Safety Education Resource Directory*, U.S. Fire Administration.
*Public Fire Education Planning, a Five-Step Process*, U.S. Fire Administration, 2002.
*Short Guide for Evaluating Local Public Fire Education Programs*, U.S. Fire Administration, 1991.
U.S. Fire Administration at www.usfa.fema.gov.

## NOTES

1. *Fire in the United States*, 12th ed. (Washington, DC: National Fire Data Center, U.S. Fire Administration, 2007), page 36.
2. *Safeguarding the Home Against Fire, A Fire Prevention Manual for the School Children of America* (New York: National Board of Fire Underwriters, 1918), page 3.
3. Percy Bugbee, *Men Against Fire, The Story of the National Fire Protection Association* (Boston: National Fire Protection Association, 1971), page 5.
4. Ibid., page 4.
5. Ibid., page 6.
6. Harry Chase Brearley, *Fifty Years of Civilizing Force* (New York: Frederick A. Stokes, 1916), page 167.
7. Ibid., page 168.
8. Ibid.
9. Charles C. Hawkins, *Fire Prevention Education* (New York: National Board of Fire Underwriters, 1942), page 11.
10. Ibid., page 12.
11. Ibid., page 11.
12. *Pioneers of Progress* (New York: National Board of Fire Underwriters, 1941), page 129.
13. *Safeguarding the Home Against Fire*, page 9.
14. Ibid., page 8.
15. *Fire in the United States*, 15th ed., page 37.
16. National Fire Academy, Fire Prevention Organization and Management Course Guide (Emmitsburg, MD: U.S. Fire Administration), pages 10–15.
17. Hawkins, page 343.
18. *Public Fire Education Planning, a Five Step Process* (Emmitsburg, MD: U.S. Fire Administration, 1979), page 2.
19. *America Burning* (Washington, D.C.: National Commission on Fire Prevention and Control, 1973), page 170.
20. Ibid.
21. *International Fire Code*, 2009 edition (Falls Church, VA: International Code Council, 2009), Section 404.
22. *World Trade Center Bombing: Report and Analysis* (Emmitsburg, MD: United States Fire Administration, 1993), page 130.
23. Ibid., page 6.
24. Ibid.
25. Rita F. Fahey and Guylene Proulx, "Study of Human Behavior During the World Trade Center Evacuation," *NFPA Journal*, March/April 1995, page 66.
26. *World Trade Center Building Performance Study* (Washington, D.C., Federal Emergency Management Agency, 2002), page 1.
27. Fahey and Proulx, page 66.

# 11

# Fire Prevention Records and Record Keeping

## LEARNING OBJECTIVES

Upon completion of this chapter, you should be able to:

- Describe what is meant by the terms *public record* and *retention schedule*.
- Describe the Freedom of Information Act (FOIA) and its impact on the day-to-day working of government.
- Discuss the reasons for the exemptions included in the FOIA.
- Discuss the benefits of electronic information management systems (IMS) in the operation of a fire prevention program.
- Discuss the potential impact of poor planning and implementation of an electronic IMS.

# LEGAL REQUIREMENTS FOR RECORD KEEPING

Similar to records of any government agency or entity, records generated by and in the possession of fire prevention bureaus are **public records**. The term *public records* is a legal term used by the federal and state governments and does not necessarily mean they are open for public view. Within the federal and state legal systems, there exists a complex set of laws and regulations regarding the management, retention, disclosure, and destruction of public records (Figure 11-1).

**public record**  a record, memorial of some act or transaction, written evidence of something done, or document considered as either concerning or interesting the public or open to public inspection; includes all documents: papers, letters, maps, books, photographs, films, sound recordings, magnetic or other tapes, electronic data-processing records, artifacts or other documentary material, regardless of physical form or characteristics.

How do these laws and regulations affect the typical fire prevention program? Better questions might be how should they impact it, and what is the potential impact if the laws and regulations are disregarded? First, these rules should guide agencies in the development of their records management systems. The agency's day-to-day operations must reflect the existence of these laws and regulations. The potential impact is almost too far reaching to contemplate. Government document managers who fail to follow prescribed rules could be disciplined or even face civil actions. Public employees whose malicious actions fall outside the regulations could be criminally prosecuted and face incarceration.

**FIGURE 11-1**
Records management is one of the most vital aspects of any fire prevention program.

So just what are these public records lurking behind the scene within the fire prevention bureau that can lead to so much trouble? Although political subdivisions might differ slightly, the common legal definition of public record is:

> A record, memorial of some act or transaction, written evidence of something done, or document, considered as either concerning or interesting the public, or open to public inspection. Any "writing" prepared, owned, used or retained by any agency in pursuance of law or in connection with the transaction of public business; and, "writings" means all documents, papers, letters, maps, books, photographs, films, sound recordings, magnetic or other tapes, electronic data-processing records, artifacts or other documentary material, regardless of physical form or characteristics.[1]

Inspection reports, notices of violation, plan review letters, permit applications, and related correspondence, whether in paper or electronic format, all fall under the public records definition. For the most part, these are pretty obvious. Other materials that are not so obvious are things including bureau phone logs, inspector's field notes and daily logs, inspections photographs, third-party inspection reports, and fuel or materials inventories submitted by businesses and in the possession of the agency, even if they are not the agency's property. Things such as fire investigation reports, witness statements, fire scene photographs, and autopsy reports are also public records, although they are not all necessarily available to the general public.

**NOTE** Fire investigation reports, witness statements, fire scene photographs, and autopsy reports are also public records, although they are not all necessarily available to the general public.

The federal laws regarding public records within the federal government are in Title 44 of the United States Code. Each state has statutes regarding records management by the state government and the political subdivisions within the state. Regardless of whether a fire prevention program is operated as a governmental function or by a private concern, the laws regarding public records impact the program. Records are generated by or for the government organization or are submitted under some requirement to the government agency. That makes the records public records. That means that everything regarding the record's birth, life, and eventual destruction should be in accordance with the appropriate law or regulation.

## Freedom of Information and Public Access Laws

**freedom of information and public access laws** federal or state laws that provide for public access to government documents and meetings.

*A popular Government without popular information or the means of acquiring it, is but a Prologue to a Farce or a Tragedy or perhaps both. Knowledge will forever govern ignorance, and the people who mean to be their own Governors, must arm themselves with the power that knowledge gives.*

—James Madison

The preceding statement from the fourth president of the United States and "Father of the Constitution" was used as the introduction to the *DOD Freedom of Information Handbook*[2] distributed by the Defense Department's (DOD's) Directorate for Freedom of Information. Our nation is at war with foreign governments and international terrorists, yet the DOD distributes a user-friendly 10-page booklet on how to request information from the military.

There is a federal law that requires the DOD to release certain information; it is probably not due to our military leaders' abiding devotion to the public's right to know. I mention this in case you think that the public safety veil might somehow exempt the fire department or fire prevention bureau from releasing public information. Within certain limitations, every government agency and entity is required by law to make information concerning the workings of government available to the people.

**NOTE** Within certain limitations, every government agency and entity is required by law to make information concerning the workings of government available to the people.

In 1967, President Lyndon Johnson signed the *Freedom of Information Act* (FOIA) passed by Congress, making it the law of the land. The law was intended to open the workings of government to the public and was the first law that gave Americans access to the records of federal agencies. Most Americans did not rush to Washington, DC, to rifle through file cabinets, but press access to agencies and their information was significantly increased. The states followed with their own statutes that ensured access to records and prohibited deliberation by public officials except in duly advertised public meetings.

## Exceptions

There are exceptions to the federal and state laws. Generally, ongoing criminal investigations, medical records, personnel records, items affecting national security, trade secrets, and attorney–client communications are exempt from disclosure. Many state statutes have been amended in the wake of the events of September 11, 2001, to include critical infrastructure to the list of exemptions.

**NOTE** Generally, ongoing criminal investigations, medical records, personnel records, items affecting national security, trade secrets, and attorney–client communications are exempt from disclosure.

Copyrighted materials such as codebooks, fire protection plans, and fire protection directories or installation manuals may be accessible for viewing, but photocopying is prohibited. **Trade secrets** must also be protected. If the formula for Coca-Cola is somehow part of a materials inventory statement submitted to ensure fire code compliance, the fire prevention bureau is responsible for ensuring that it remains a secret.

Allowing trade secrets to be released and potentially used by business competitors is illegal and unethical.

| **trade secrets** | propriety information regarding a business or manufacturing process. |

## Practical Application

How do these laws potentially affect those in the fire prevention business? For government employees, the impact is extensive. Jurisdictions need to have policies that ensure compliance while maintaining necessary safeguards. Individual employees should never freelance or attempt to make legal determinations as to whether exemptions exist. When in doubt, the jurisdiction's legal representative should be immediately contacted.

**NOTE** Jurisdictions need to have policies that ensure compliance while maintaining necessary safeguards.

**NOTE** When in doubt, the jurisdiction's legal representative should be immediately contacted.

Meetings of the governing body and appointed boards and commissions must also comply. If the fire code appeals board is scheduled to hear an appeal at 7:00 PM, it is illegal for the board members to discuss the matter in the parking lot while entering the building at 6:45. Discussions must take place in public at the appointed time, where all interested parties have the opportunity to observe the proceedings. It is an unfortunate fact that government business was sometimes conducted in secret in the past in order to exclude certain groups. Freedom of Information or Sunshine laws were passed to end the unseemly practice.

## RECORD RETENTION AND STORAGE

The maintenance of records by fire prevention bureaus is generally governed by regulations promulgated under state statutes by a state entity that serves as the state archivist. Regulations for the retention and storage of records of fire departments and fire prevention bureaus within the State of Texas can be found in *Local Schedule PS, Retention Schedule for Records of Public Safety Agencies*[3] (Figures 11-2 and 11-3). Failure to retain the records in accordance with the regulations is a criminal act.

**NOTE** Failure to retain the records in accordance with the regulations is a criminal act.

## RECORDS MANAGEMENT SYSTEMS

Records of activities are organized or gathered into some type of system. The stack of business cards in your desk drawer that you have not had the time to put into alphabetical order is actually an information management system (IMS). It is not a very efficient

**LOCAL SCHEDULE PS (2nd edition)**

**Retention Schedule for Records of Public Safety Agencies**

*Effective October 20, 1997*

This schedule establishes mandatory minimum retention periods for the records listed. No local government office may dispose of a record listed in this schedule prior to the expiration of its retention period. A records control schedule of a local government may not set a retention period for a record that is less than that established for the record on this schedule. The originals of records listed in this schedule may be disposed of prior to the expiration of the stated minimum retention period if they have been microfilmed or electronically stored pursuant to the provisions of the Local Government Code, Chapter 204 or Chapter 205, as applicable, and rules of the Texas State Library and Archives Commission adopted under authority of those chapters. Actual disposal of such records by a local government or an elective county office is subject to the policies and procedures of its records management program.

Destruction of local government records contrary to the provisions of the Local Government Records Act of 1989 and administrative rules adopted under its authority, including this schedule, is a Class A misdemeanor and, under certain circumstances, a third degree felony (Penal Code, Section 37.10). Anyone destroying local government records without legal authorization may also be subject to criminal penalties and fines under the Open Records Act (Government Code, Chapter 552).

**INTRODUCTION**

The Government Code, Section 441.158, provides that the Texas State Library and Archives Commission shall issue records retention schedules for each type of local government, including a schedule for records common to all types of local government. The law provides further that each schedule must state the retention period prescribed by federal or state law, rule of court, or regulation for a record for which a period is prescribed; and prescribe retention periods for all other records, which periods have the same effect as if prescribed by law after the records retention schedule is adopted as a rule of the commission.

Local Schedule PS sets mandatory minimum retention periods for records series (identified in the Records Series Title column) commonly found in law enforcement agencies, fire departments and rural fire prevention districts, emergency medical departments, emergency communications agencies and districts, county medical examiner departments, county and district attorneys offices, and community supervision and corrections departments. In addition to counties and cities, this schedule should also be used by other local governments, such as junior college

**FIGURE 11-2**

Records retention and maintenance is generally regulated by state archivists.

4450-03 **FIRE RECORD** - A log, register, consolidated daily or other periodic report, or any other form of record that provides in summary form information on each fire or other incident to which fire or emergency medical personnel have responded, including at a minimum the date, time, location, and nature of the incident. RETENTION: 2 years; or 2 years after last entry if in bound volume.

4450-04 **INCIDENT REPORTS** - Reports, including those completed on Texfirs or other incident reporting system forms, of each fire or other incident to which a fire fighting or other fire agency unit has responded, detailing the type of incident, units responding, action taken, equipment used, and other pertinent data. RETENTION: 5 years.

### SECTION 4-2: FIRE PREVENTION AND INSPECTION RECORDS

4475-01 **ALARM PERMITS AND ASSOCIATED DOCUMENTATION**

   a) Installation certificates for fire detection and fire alarm devices or systems filed with fire agencies by rule (37 TAC 531.17) of the Texas Commission on Fire Protection. RETENTION: Life of device or system.

   b) Applications for fire detection and alarm permits, copies of permits or other documentation evidencing issuance, and any inspection or evaluation reports prepared during a permit period, if permits are required by local policy. RETENTION: Expiration or revocation of permit + 3 years for granted permits; date of denial + 1 year for denied permits. (Documentation on denied permits is exempt from destruction request to the Texas State Library.)

4475-02 **AUTOMATIC SPRINKLER SYSTEM PERMITS AND ASSOCIATED DOCUMENTATION**

   a) Automatic sprinkler material and test certificates filed with fire agencies by rule (37 TAC 541.16) of the Texas Commission on Fire Protection. RETENTION: Life of system.

   b) Applications for automatic sprinkler system permits, copies of permits or other documentation evidencing issuance, and any inspection or evaluation reports prepared during a permit period, if permits are required by local policy. RETENTION: Expiration or revocation of permit + 3 years for granted permits; date of denial + 1 year for denied permits. (Documentation on denied permits is exempt from destruction request to the Texas State Library.)

* 4475-03 **CERTIFICATES OF OCCUPANCY** - Copies of certificates of occupancy or record of their issuance used to certify final approval for the occupancy of new structures or old structures that have been remodeled to the extent that a certificate of occupancy is required by local

**FIGURE 11-2**
continued

**FIGURE 11-3**

Retention schedules developed by state archivists stipulate time frames for records storage and retention.

| RECORD | RETENTION |
|---|---|
| Installation certificates for detection and alarm devices | Life of the system |
| Alarm and detection system permits | 3 years for approved permits<br>1 year for rejected permits |
| Sprinkler system test and material certificates | Life of the system |
| Sprinkler system permits | 3 years for approved permits<br>1 year for rejected permits |
| Certificates of occupancy | As long as administratively valuable |
| Complaints | 3 years from resolution of the complaint |
| Controlled burn records | 1 year |
| Drill and simulation records | 5 years |
| Hazardous materials permits | 3 years for approved permits<br>1 year for rejected permits |
| Inspection reports and logs | 3 years if master record for the structure is maintained for the life of the building |
| Notice of violation | 3 years from correction of the violation |
| Plan review records | As long as administratively valuable |

IMS because you have to manually look at every card before you find the one you want, but it is a system nonetheless. Your IMS for business contacts is a random stack of business cards tucked within your desk.

Effective records management is essential for an effective fire prevention program. Records of inspections, plan reviews, and investigations are the basis for criminal and civil actions. Cases can be won or lost based on a single report or perhaps on the inability to find a single report. In addition to legal issues regarding fire inspection and investigation records, there are the business issues.

**NOTE** Cases can be won or lost based on a single report or perhaps on the inability to find a single report.

Fire prevention bureaus and state fire marshal offices are subgroups or sections within larger agencies, usually a part of government. Governments may not be businesses in the true sense, but some business issues are fundamental to the success and survival of organizations and their missions.

Terms such as *budget, funding, expenditures, cost recovery, management indicators,* and *productivity* are not found in most firefighting texts, but skill in performing the "business" functions of government is every bit as important to the success of the mission as preventing or extinguishing fires. You cannot perform many inspections without vehicles and fuel. Hoses carried on fire apparatus have to be purchased, but they must be justified within a budget first. To justify funding and expenditures, statistics for various functions performed by the unit are vital. The same statistics are used to measure the effectiveness of the fire prevention program, identify trends, and compare the jurisdiction's fire experience with that of other localities and the nation.

Information on inspections, investigations, plan review, and public education activities should be tracked and reported by employees, by work groups, and for the bureau as a whole. Care should be taken to identify useful information elements that are clearly defined and easily tracked. If inspections are tracked by type using terms such as *routine*, *follow-up*, and *special*, every staff member needs to understand the meaning of the terms and classifications. Finding out that half of the staff had a slightly different understanding of what constituted a "special" inspection could render months or years of statistics virtually worthless.

Data regarding employee productivity, number and types of inspections, investigations, plan review activities, and public education programs can be used to justify additional staff and resources and track employee performance and can even be used as a basis for cost recovery or user fees. The importance of accurate statistics of the organization's activities cannot be overstated.

An issue that frequently gets lost in the shuffle is the fact that the fire prevention bureau is part of a larger organization, normally within a government of some type. Is there value in the ability to share information with other organizational units or with other agencies? Should the fire department operations unit have access to the hazardous materials inventory records that the fire prevention bureau has collected? Would it be useful for firefighters responding to an alarm in the middle of the night to be able to readily access building information within the bureau's records system?

The potential value of sharing information is enormous. The safety of emergency responders can be enhanced, and communications within the organization and within the government can be enhanced. If a fire investigator on the scene has access to fire inspections records and building construction details, the complex process of determining the origin and cause of the fire might become a little easier. Time normally spent tracking down basic information on the structure produces significantly greater dividends by conducting additional interviews immediately after the fire.

**NOTE** The potential value of sharing information is enormous.

## Types of Records Management Systems

In addition to the legal requirements for record keeping that identify security, retention, and public access, several practical requirements must be addressed. The system used must be efficient, effective, and sustainable. The information gathered should reflect the mission of the organization. Collecting and maintaining statistics should not become the mission. This important issue should not be brushed aside. Many public and private organizations have invested considerable money and effort in complex systems designed to streamline and improve records management only to find that the system actually reduced efficiency.

**NOTE** The information gathered should reflect the mission of the organization. Collecting and maintaining statistics should not become the mission.

Records management systems range from two-drawer file cabinets to complex computer systems, with countless variations of complexity in between. But complex does not necessarily mean better. Efficiency, reliability, and ease of use are the real measures of the system. A complex electronic IMS that proves to be less reliable and more labor intensive than the manual system it replaces is a step backward—and an expensive step at that.

## Manual Systems

Most organizations maintain at least some records manually. In *Managing Fire Services*, Phillip Schaenman succinctly identified the reason that few organizations ever really go paperless. "Paper is transportable and cheap, can still be used in power failures, can be looked at in private, and does not take expertise to use."[4] The downside of paper records is that paper takes up considerable space and must be manually filed and searched. Manipulation of data must be accomplished by hand, making statistical study and comparison an arduous task at best.

Many organizations transfer paper records to film or to electronic images in an effort to conserve space. For the most part, these images are just another form of manual systems. There are electronic systems that have **optical character recognition (OCR)** capability, meaning the system "reads" written text, translates it into digital format, and stores it as a digital record. The penmanship skills of the staff become critical when machines are used to read and translate written documents.

**optical character recognition (OCR)** software that "reads" written documents when scanned and digitizes characters.

## Electronic Information Management Systems

Computers can imitate a manual system, basically taking the same filing system and storing it digitally. It takes up less space, and the data can be sorted, searched, and manipulated by input fields such as date or address or business name. The more input fields used, the broader the capability of data manipulation. Unfortunately, each data field input requires a human action, such as keying in an address, selecting a type of occupancy from a list, or entering some other code. There is a fine line between useful data collection and useless information overload.

Computerized IMSs that integrate records storage with the processes that generate records are a step up in complexity. Inspections programs that feature handheld PDAs (personal data assistants) with the ability to upload data collected in the field and generate reports are available and in use in some jurisdictions. Building information data covering every construction detail can be stored in a small hologram and applied to the main electrical panel cover. On arrival at an inspection site, the inspector simply scans the sticker with the PDA, and an inspection form complete with building information and construction details appears on the PDA touch screen. The inspector begins the inspection using a series of code-based checklists. At the end of the day, information can be uploaded to the bureau's computer by telephone modem or directly using infrared (IR) technology.

Such systems must be written for the code adopted by the jurisdiction, have the capacity to be modified to reflect code updates, and meet the legal requirements for reporting the inspection. Most fire codes require that a written notice of violation be issued each time a violation is encountered. In most cases, the notice must include the unsafe condition, applicable code section, required corrective action, and date of the follow-up inspection. If the system cannot meet the legal requirements set out for the inspection process, it is worse than useless.

Some software developers either do not have an adequate grasp of the technical needs of a fire prevention bureau or they use a consultant with fire department experience but without fire prevention bureau experience. Either way, an unusable system can be developed or procured. In some cases, fire inspection or investigation may simply be an add-on module to a program that manages shift scheduling, training records, and other fire department functions. Such systems can do a good job in many instances with adequate planning, coordination, and effort. In any case, a clear understanding of the bureau's operational needs must be clearly spelled out before a contract is negotiated.

Preliminary fire investigation reports can also be developed for PDAs, allowing investigators to scan building information in the course of investigating commercial building fires. Information for structures without stickers would simply be manually entered.

Technology is a two-edged sword. The potential for increasing efficiency and productivity is very real. Without proper planning and implementation, going high tech can quickly derail any activity, and fire prevention programs are not exempt from the possibility. Before taking the big plunge into an electronic IMS, a detailed analysis of the organization's operational needs must be performed and compared with the capabilities of the proposed system. Among the issues that must be considered are the following:

- The system must conform to the legal requirements of the jurisdiction for public records.
- If the system incorporates inspection, investigation, or other code or law enforcement elements, do they meet the requirements of the jurisdiction?
- The system should meet the needs of the organization and be adaptable to normal operating procedures. Changing effective operating procedures to compensate for the shortcomings of a system that was supposed to make things more efficient is evidence of poor planning or poor communication between the organization and the vendor.

- A "smart" system will not compensate for poorly trained personnel. The notion that technology reduces the need for technical competence is a dangerous one. A poorly trained inspector or investigator equipped with the best technology is still poorly trained and not up to the job.

- Technical support, the availability of future software updates, and integration with other systems are every bit as important as the system capability.

- What is the life expectancy of the media used for data storage? Optical media such as CD-ROM and WORM (write once, read many) have an estimated life of about 30 years.[5] What steps will be taken to preserve documents that must be maintained for longer periods of time?

- The more time spent up front identifying the needs of the organization, carefully selecting the system with the best fit for the organization's needs, testing the system before going online, and training the users, the greater the chance for success.

**NOTE** Without proper planning and implementation, going high tech can quickly derail any activity, and fire prevention programs are not exempt from the possibility.

## FINANCIAL RECORDS

Financial records associated with fire prevention programs are generally a very small aspect of the records system, although some fire prevention bureaus generate large sums through the use of user fees. Whether it is a $2 fee for copying a report or several thousand dollars in permit and plan review fees for the construction of a high-rise building, financial records must be generated and maintained in accordance with the appropriate regulations.

Chapter 13, "Financial Management," provides an in-depth discussion of fee and cost recovery systems, but a discussion regarding the records associated with the process rightly belongs in this chapter. In addition to state laws regarding the financial records of government agencies, the system used by the fire prevention bureau should conform to the system in use by its parent government organization. Fire prevention bureaus within city fire departments should consult with their city finance department counterparts for guidance. There is a good chance that the jurisdiction has an accounting system in use that can be immediately placed in service, conforms to state requirements, and will automatically generate required reports.

The notion that funds could actually be embezzled in a fire prevention bureau might seem remote, but cases of misappropriation of funds have occurred in every type of organization from churches and orphanages to law enforcement agencies. Financial records management systems are designed to keep honest people honest and facilitate the audit process.

**NOTE** Financial records management systems are designed to keep honest people honest and facilitate the audit process.

## SUMMARY

Records generated by and in the custody of fire prevention bureaus are public records and are regulated by state or federal laws and regulations. Retention schedules, generally developed by state archivists, contain requirements that must be followed regarding the retention, storage, and possible destruction of government documents.

Records are maintained within some type of IMS. Whether the system is a manual paper system, a fully computerized system, or something in between, it must conform to the regulations regarding public records.

The FOIA, a federal law passed in the 1960s, is designed to open the workings of government. States adopted similar legislation that affected the states and their political subdivisions. Paper and electronic records, photographs, maps, recordings, and texts are all considered public records, and unless exempted from disclosure, must be made available for public inspection. The same laws prohibit deliberation by public bodies except at duly authorized public meetings unless specifically exempted.

IMS must be developed and implemented to complement the organization's normal operating procedure, thereby improving efficiency. Operating procedures should not have to be radically altered to accommodate an IMS.

## REVIEW QUESTIONS

1. What is the term generally used for state regulations that specify time periods for the storage and disposal of public records?
2. Do Freedom of Information and Public Access laws apply to noncompensated citizen boards such as a fire code appeals board?
3. List four items that are generally exempt from disclosure under the FOIA.
4. List two instances in which access to fire inspection information could benefit other units within a fire department.
5. List four issues to be considered in evaluating a new IMS.

## DISCUSSION QUESTIONS

1. You have been called to the fire chief's office to attend a sales presentation of an IMS featuring handheld computers for use in the field by fire inspectors. The system has already been purchased by the building and zoning department for use in managing construction inspections. The chief does not want to fall behind the other agencies in the local government's efforts to embrace technology. What are some of the issues you will ask the salesperson to address during the presentation?

2. A citizen has requested a copy of the jurisdiction's fire code. When told by a staff member that the code was the *International Fire Code* and could be purchased, the citizen requested a photocopy. When his request was refused, he demanded a copy under the FOIA. Is the citizen correct? How can this issue be resolved?

## CHAPTER PROJECT

Using your jurisdiction's FOIA law, determine whether the following must be released to the public:

- The fire chief's salary
- The fire marshal's most recent performance evaluation
- Citizen complaints regarding unsafe conditions at a nightclub
- Sprinkler plans for the jurisdiction's wastewater treatment plant
- Sprinkler plan for an existing restaurant
- Report of inspection for a local chemical plant
- Building diagrams for the chemical plant
- List of hazardous materials stored at the National Guard Armory

## ADDITIONAL RESOURCES

In-depth information on many of the subjects discussed in this chapter can be found in the following texts and publications and at these Web sites.

FOIA handbooks for most states are available online or in local libraries.

National Technical Information Service (NTIS), *Department of Defense FOIA Handbook* or available online at http://www.dod.mil/pubs/foi/docs/FOIAhandbook.pdf

## NOTES

1. M.J. Connolly, et al., *Black's Law Dictionary*, 6th ed. (Saint Paul: West Publishing, 1991), page 882.
2. Directorate for Freedom of Information and Security Review, *DOD Freedom of Information Handbook* (Washington, DC: Department of Defense, 2003), page 1.
3. *Local Schedule PS, Retention Schedule for Records of Public Safety Agencies*, 2nd ed. (Austin: Texas State Library and Archives Commission, October 20, 1997).
4. Dennis Compton, et al., Managing Fire and Rescue Services 3rd ed. (Washington, DC: International City Managers Association, 2002), page 130.
5. *The National Archives and Records Administration and the Long-Term Usability of Optical Media for Federal Records: Three Critical Problem Areas* (Washington, DC: The National Archives and Records Administration, 2003), page 1.

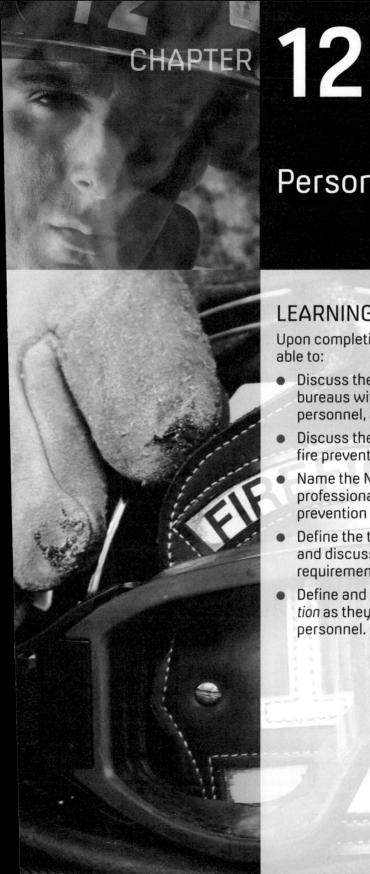

# CHAPTER 12

# Personnel

## LEARNING OBJECTIVES

Upon completion of this chapter, you should be able to:

- Discuss the benefits of staffing fire prevention bureaus with uniformed firefighters, civilian personnel, and a combination of both.

- Discuss the value of national certification for fire prevention bureau staff.

- Name the National Fire Protection Association professional qualifications standards for fire prevention bureau personnel.

- Define the term *job performance requirement* and discuss the impact of job performance requirements on certification.

- Define and contrast *accreditation* and *certification* as they apply to fire prevention bureau personnel.

# PERSONNEL

There is an old joke about a fire chief who is asked during his retirement ceremony to reflect on his 30 years in the fire department. "All in all, it was a great job, except for the people." People are the job. People are the organization. There is nothing more important to the success of the job of protecting the public than selecting and nurturing the best people for the job.

**NOTE** There is nothing more important to the success of the job of protecting the public than selecting and nurturing the best people for the job.

Traditionally, fire prevention duties within fire departments were performed by firefighters. In many cases, assignment to the bureau of fire prevention was considered a promotion and included an increase in pay. Today, fire prevention bureaus are often made up of a combination of **uniformed** and **civilian employees**, the former being firefighters and the latter being specialists who possess specific skills, education, or training. Some bureaus have shifted entirely away from uniformed firefighters. There are specific advantages and some disadvantages to each method of staffing.

**uniformed employees** trained firefighters within a fire department.

**civilian employees** fire department employees in a support position who are not trained firefighters.

## Uniformed Firefighters

The traditional method of staffing fire prevention bureaus with uniformed personnel has the distinct advantage of using personnel with a demonstrated fire service background and a known record of work experience (Figure 12-1). The skills needed to be a successful investigator, inspector, plans examiner, or public fire educator are not necessarily those that make a good firefighter. Transfer to the bureau may require the acquisition of new, very different skills.

**NOTE** The traditional method of staffing fire prevention bureaus with uniformed personnel has the distinct advantage of using personnel with a demonstrated fire service background and a known record of work experience.

Many in the fire service have their favorite tale regarding a presentation at a school or public gathering in which a firefighter's ill-chosen terminology or impatience with children or clueless adults created a bit of a stir. Usually the offense was harmless, embarrassing the officer in charge and forcing the rest of the crew to hide their smiles to escape the wrath of the officer on the trip back to quarters. Many firefighters simply are not interested in the changes that an assignment to the fire prevention bureau sometimes causes. Fire prevention assignments usually entail a transfer from shift to weekdays with a loss of considerable free time and may preclude the opportunity for outside part-time

**FIGURE 12-1**
Fire prevention
bureaus have
traditionally been
staffed with uniformed
firefighters. (Courtesy of
Jeremy Luttrell.)

employment. Many firefighters simply cannot afford the loss of shift differential, holiday pay, and outside income.

For others, the move from working as part of a tightly knit team with constant interaction to the mostly solitary job of inspector or investigator is too great a change (Figure 12-2). Others welcome it as a needed challenge after too many years on the floor. For those who are interested in a change, assignment to the bureau can be the challenge that recharges their batteries and energizes them in their fire service career. Too often, we tend to assess the advantages and disadvantages regarding personnel issues from only one viewpoint. Some focus on the well-being and happiness of the individual, disregarding the fundamental fact that service and the welfare and safety of the public are the missions of the organization. Others dismiss the legitimate needs and desires of the employees, forgetting that the organization's only legitimate assets are its people and that attitudes affect performance.

**NOTE** Service and the welfare and safety of the public are the missions of the organization.

An issue frequently cited as a disadvantage of staffing the bureau with uniformed firefighters is that many undergo extensive training, attain certification, and become proficient in their position, only to be promoted out of the bureau. The issue is a real concern for managers of fire prevention programs. The cost of training and the difficulty in recruiting and retaining skilled personnel are issues in any organization. What may

**FIGURE 12-2**
Retired firefighters make excellent civilian inspectors due to their in-depth knowledge of fire prevention and protection and years of street experience.

be a disadvantage for the bureau is actually a great benefit to the department and to the public. Chief officers, who have gained street-level experience as investigators, inspectors, plans examiners, and fire safety educators, have the benefit of their experiences in dealing with the public, the business community, and the criminal justice system. The fire prevention bureau is considered "finishing school" for chief officers by some departments.

**NOTE** Chief officers, who have gained street-level experience as investigators, inspectors, plans examiners, and fire safety educators, have the benefit of their experiences in dealing with the public, the business community, and the criminal justice system.

## Light-Duty Assignments

Firefighting is a rigorous, dangerous, and physically and mentally demanding job. Some positions within the bureau are suitable for firefighters who can no longer perform the activities of a firefighter. However, great care must be taken to ensure that the candidate can perform all the necessary tasks for the position in the bureau. Inspectors literally walk miles every day, much of it up and down stairs or ladders. Fire investigators work in fire scenes under the worst possible conditions, and some are law enforcement officers who must be capable of physically defending themselves, subduing suspects, and operating at all hours and under the worst extremes of temperature.

# Civilian Employees

Many fire departments that established plans review programs quickly realized that training and retaining competent plans examiners using uniformed firefighters was extremely difficult. The number of plans examiner positions within most bureaus was generally so small that the loss of a single examiner threw the program into chaos. The time lag between appointment and the employee's attaining the level of technical skill needed to do the job was so long that it threatened the bureau's ability to provide the service. Delays in plan review and approval are unacceptable to developers, builders, contractors, and politicians, and rightly so.

Some fire departments established civilian positions within the bureau for some or all of the plans examiner positions. Registered fire protection engineers routinely review plans and supervise plans review branches in some bureaus. Many fire departments hire graduates from engineering schools, giving them valuable public sector experience and, hopefully, selling them on the importance of fire prevention and protection during their first years as engineers.

By establishing positions for civilian fire protection engineers, fire prevention bureaus gained both the stability of long-term employees and the technical expertise of engineering school graduates. The relationship is mutually beneficial in that engineers can call on firefighters for guidance regarding tactical considerations such as firefighting access, fire protection equipment, and fire behavior.

**NOTE** By establishing positions for civilian fire protection engineers, fire prevention bureaus gained both the stability of long-term employees and the technical expertise of engineering school graduates.

Many departments have hired teachers as fire safety educators, former law enforcement officers as investigators, and construction inspectors and fire protection contractors as inspectors. These specialists bring a breath of fresh air to the organization with new ideas and skills. However, training must be provided in fire department procedures and methods and even fire department history and culture. Many fire departments have successfully integrated civilian positions into their fire prevention bureaus with great success. In some cities and counties, employees from other agencies flood the fire department with applications when vacancies for civilian positions within the bureau are announced.

Another advantage of establishing civilian positions within the bureau is economic. Less turnover means fewer dollars in training costs. Comparisons of fringe benefit packages for uniformed firefighters and civilian employees (uniformed retirement, workers' compensation, and insurance) typically reveal differences of more than 10 percent. It is generally cheaper to establish civilian positions. Public safety employee positions (fire and police) have higher associated personnel costs.

The big disadvantage of reclassifying uniformed positions into civilian positions is that it reduces the number of opportunities for firefighters to gain the valuable experience of working in the fire prevention bureau. Probably the best system is one that balances the needs of the fire prevention bureau for stability and consistency in key positions and the need of the department for a pool of well-rounded chief officer candidates. The experience gained in the bureau more closely represents the interaction with business and political leaders that makes up a great part of a chief officer's job.

**NOTE** The big disadvantage of reclassifying uniformed positions into civilian positions is that it reduces the number of opportunities for firefighters to gain the valuable experience of working in the fire prevention bureau.

**NOTE** The experience gained in the bureau more closely represents the interaction with business and political leaders that makes up a great part of a chief officer's job.

The terms *uniformed* and *civilian* as they apply to positions within the fire department do not have much to do with clothing, uniforms, and methods of identification. To the public, a civilian inspector who is a fire department employee is the fire department. There are important issues of public perception as well as significant safety issues. Civilian employees are not firefighters, they are not trained as emergency responders, and standard operating procedures must clearly state when and under what circumstances they are permitted to enter the fire scene or approach hazardous locations. Employees must not be exposed to hazards beyond their level of training.

Assigning untrained civilian personnel to marked fire department sedans potentially exposes them to emergency situations on the road. When an untrained employee is flagged down by distraught bystanders at an accident scene, the employee and the department are caught in a no-win situation. If the employee refuses to act beyond his or her level of training, cries of negligence will be heard from those who think that every person who drives a red and white sedan must be a firefighter or emergency services technician. The other possibility is that an employee lacking training and personal protective equipment is exposed to any number of potentially fatal hazards for trying to be a Good Samaritan. There are marking schemes that eliminate or at least significantly reduce this risk.

The issue of uniform clothing issue or rules and regulations regarding dress is also an important one. Fire departments are paramilitary organizations, and most firefighters are proud of their department and proud of their positions as firefighters. When civilian employees are issued the identical uniform, down to the collar pins, many firefighters will rightly feel slighted. The firefighters attained their position through competitive exams, graduated from the academy, and believe that wearing the uniform should be reserved for those who have met the same challenges.

Many fire departments have adopted uniforms for civilian employees that are slightly different from those worn by firefighters. These provide a consistent appearance when in the public view, make civilian employees feel like they are really part of the fire department, and keep the uniformed firefighters secure in the knowledge that their professional status is intact.

Security has taken its rightful place in society. For too long, individuals who have claimed to be public officials or representatives of public utility companies have been given almost unlimited access to American homes and businesses. Fire prevention bureau employees must be provided with credentials that include photo identification. Standard operating procedures should require investigators or inspectors to present their identification, not just a badge, every time they seek admission to any premises or conduct interviews.

**NOTE** Standard operating procedures should require investigators or inspectors to present their identification, not just a badge, every time they seek admission to any premises.

## TRAINING AND CERTIFICATION

Many, if not most, of the duties assigned to the members of the fire prevention bureau require job-specific training and, in many cases, certification. Most of the skills required for the various jobs are not skills that firefighters receive as part of their standard training (Figure 12-3). Many states have training and certification requirements

**FIGURE 12-3**
Training and job-related skills for fire investigators can be quite different from those for firefighters. (Courtesy of Duane Perry.)

for fire investigators, fire inspectors, and plans examiners that mandate initial and refresher training and certification through testing. National certification programs have proven to be of real benefit to the fire service and fire protection and prevention industry.

## Standards for Professional Qualifications

Personnel issues that include upward mobility, salaries, fringe benefits, light-duty assignments, technical expertise and retention of technical skills, and consistency of service all must be considered in establishing and funding positions. A valuable tool in establishing positions, developing **KSAs (knowledge, skills, and abilities)** advertising and hiring, and determining base salaries that are in line with jobs with comparable skill levels are the National Fire Protection Association (NFPA)'s professional qualifications standards. The NFPA maintains standards for professional qualifications for fire service personnel, including traditional positions within fire prevention bureaus:

- NFPA 1031: *Standard for Professional Qualifications for Fire Inspector and Plan Examiner*
- NFPA 1033: *Standard for Professional Qualifications for Fire Investigator*
- NFPA 1035: *Standard for Professional Qualifications for Public Fire and Life Safety Educator*

**KSAs (knowledge, skills, and abilities)** minimum knowledge and attendant skills and ability needed to perform a job.

The need for nationally recognized performance standards for fire department personnel became readily apparent to fire service leaders in the 1960s and 1970s. The National Board on Fire Service Professional Qualifications (Pro Board) for the fire service was established in 1972 to facilitate the development of job performance standards for uniformed fire service personnel. The "Pro Board," as it is commonly called, initially established four technical committees to develop standards for firefighter, fire officer, fire service instructor, and fire inspector and investigator using the NFPA's standards-making system.[1]

**NOTE** The need for nationally recognized performance standards for fire department personnel became readily apparent to fire service leaders in the 1960s and 1970s.

The NFPA's professional qualification standards are used extensively throughout the United States and are used as the basis for certification of personnel. The U.S. Department of Defense (DOD) adopted the certification system for DOD fire departments worldwide, making it the largest group of nationally certified firefighters in the world.

The NFPA describes the benefits derived from the system of establishing professional qualifications to enable voluntary certification:

> The primary benefit of establishing national professional qualification standards is to provide both public and private sectors with a framework of the job requirements for the fire service. Other benefits include enhancement of the profession, individual as well as organizational growth and development, and standardization of practices. NFPA professional qualification standards identify the minimum **JPRs (job performance requirements)** for specific fire service positions. The standards can be used for training design and evaluation, certification, measuring and critiquing on-the-job performance, defining hiring practices, and setting organizational policies, procedures, and goals. (Other applications are encouraged.)[2]

**JPRs (job performance requirements)** a statement within a certification standard that describes a specific job task, lists the items necessary to complete the task, and defines measurable or observable outcomes and evaluation areas.

The standards incorporate a system of JPRs in which skills are grouped by major areas of responsibility. For each JPR, three components are listed in the standard: (1) a skill; (2) materials, tools, or information required to accomplish the task; and (3) an evaluation parameter or desired outcome. Each JPR also includes requisite knowledge needed for the candidate to accomplish the task. Inspector I, II, and III JPRs are organized into three categories: administration, field inspection, and plan review.

There are three levels for fire inspector: Fire Inspector I (entry-level personnel who conduct basic fire inspections), Fire Inspector II (journeyman or intermediate-level inspector), and Fire Inspector III (most advanced level who performs all types of fire inspections, plans review duties, and resolves complex code-related issues). The ability to accurately calculate occupant loads is a necessary skill for fire inspectors. The JPRs and skills become progressively more complex as personnel progress through the levels:

| | |
|---|---|
| Fire Inspector I | *Compute the allowable occupant load of a single-use occupancy or portion thereof...*[3] |
| Fire Inspector II | *Compute the occupant load of a multiuse building...*[4] |
| Fire Inspector III | *Assess alternative methods to adjust occupant loads...*[5] |

To become an Inspector I, a candidate must demonstrate a basic knowledge of the skill. The Inspector II level requires a good working knowledge and intermediate technical skill. At Inspector III, the candidate must understand the underlying principles behind the skill in order to make complex judgments regarding technical equivalence when strict compliance is not possible.

## Certification

Certification for fire prevention personnel is mandated by some states but is always desirable. When an agency requires its employees to attain national certification, the bar for technical knowledge and skill is set at the level dictated by nationally recognized organizations. Attaining and maintaining certification demonstrates the ability of an individual to perform critical tasks necessary for competent performance in the field.

**NOTE** Certification for fire prevention personnel is mandated by some states but is always desirable.

Certification is particularly important for individuals who are called on to testify in legal proceedings. Witnesses ordinarily testify regarding facts that they have personally seen or heard. For fire investigators and inspectors to offer opinions during testimony, they must satisfy the court that they qualify as expert witnesses. While on the stand, expert witnesses are allowed to offer opinions in response to questions that assist juries in understanding complicated and technical subjects not within the understanding of the average layperson.[6]

**NOTE** Certification is particularly important for individuals who are called on to testify in legal proceedings.

## Certification and Accreditation

The terms **certification** and **accreditation** and which organizations perform each function are often confused by those seeking to become certified. *Accredit* means to authorize or acknowledge. *Certify* means to authenticate or vouch for. There are two fire service accreditation organizations in the United States, the National Board on Fire Service Professional Qualifications (Pro Board), and the International Fire Service Accreditation Congress (IFSAC). These agencies assess the fitness of organizations to certify individuals as meeting the standards through testing and evaluation.

**certification** approval by an accredited agency that an individual meets the requirements of a standard, accompanied by the issuance of a certificate.

**accreditation** the act of assessing the fitness of an organization to test and certify individuals in accordance with a standard; accredited agencies are authorized to test and certify individuals.

Each agency has established criteria intended to ensure an effective and nondiscriminatory testing and evaluation process that provides adequate security, safety, and consistency. Organizations must document and demonstrate that the criteria are diligently maintained. Certifying organizations are subject to site visits and audits by the accreditation agencies. Organizations that are accredited provide training and testing and evaluation services. When individuals successfully complete the testing and evaluation process, they become certified by the certifying organization. A certificate is then issued by or on behalf of the accreditation agency.

# Certifying Organizations

State fire service training agencies, government fire service training agencies, model code organizations, and academic institutions generally provide accredited certification programs. Many offer a package of training and testing and evaluation. Some certifications for fire prevention personnel are not specifically fire service certifications, including those from model building code organizations and organizations that specialize in engineering and fire protection.

The International Association of Arson Investigators (IAAI) and National Association of Fire Investigators (NAFI) have certifications for fire investigators as discussed in Chapter 9. The IAAI's Certified Fire Investigator (CFI) program has been adopted by the Bureau of Alcohol, Tobacco, Firearms and Explosives (ATF) to certify its agents as fire investigators. The National Institute for Certification in Engineering Technologies (NICET) offers certification for technicians in various fire protection fields, including sprinkler system layout, inspection and testing of water-based fire protection systems, and fire alarm systems.

The International Code Council (ICC) offers certification for fire inspectors and plans examiners as well as other code-based certifications. Fire prevention bureau personnel who become certified in code disciplines beyond those normally required, such as Certified Building Official (CBO) and Certified Professional Code Administrator (CPCA), demonstrate their desire to be on par with their counterparts in building departments. This helps to establish mutual respect and makes for better working relationships among agencies that have at times acted more like rivals than allies.

# PERSONNEL RETENTION AND ADVANCEMENT

There is nothing more important to the success of a fire prevention program than the men and women who walk through the front door of the building every morning. Personnel are a precious resource that must be nurtured and supported and then given the freedom to perform the job. The best decision is one made on the street, closest to the information. A manager who tries to call the shots from the office, based on information filtered through the very staff members who lack his or her confidence in the first place, makes bad decisions.

**NOTE** There is nothing more important to the success of a fire prevention program than the men and women who walk through the front door of the building every morning.

Well-trained and confident employees are a great asset to a fire prevention program. Technical skills that may take years to hone cannot be simply taught or gained from instruction. One of the challenges faced by most fire prevention bureaus is the retention of their best and brightest. In many cases, the only way for firefighters to be promoted and advance through the ranks is to leave the bureau. In some cases, employees find a niche they love and simply stop taking promotional examinations. Other times they must make the painful decision of leaving a job they love to better provide for their family and future.

In the big picture, the department as a whole is much better off with a large number of potential chief officer candidates who have worked in several areas of the department. For fire prevention bureau managers, that same big picture is nice, but the immediate problem is how to recruit suitable replacements and get them up and running.

Another significant issue is that of civilian personnel who simply do not have a career ladder within the bureau or the department. Watching their firefighter counterparts compete for promotional opportunities that they lack can lead to dissatisfaction. Some creative methods have been used in fire prevention bureaus that do not necessarily fix the problem but do go a long way to demonstrate the organization's desire to provide for its employees. Monetary rewards for education, certification, or performance have been used with varying degrees of success. Training opportunities involving out-of-town travel can also go a long way to reward conscientious employees.

## SUMMARY

Fire prevention bureaus have traditionally been staffed with uniformed firefighters. The system has had the advantage of enabling fire prevention bureau managers to select employees from a known candidate pool with documented fire service experience. Many modern fire prevention bureaus have established civilian positions for plans examiners, fire inspectors and investigators, and fire safety educators that can be filled with people from outside the fire department. Although this setup limits opportunities for firefighters within the organization, it enables fire prevention bureau managers to hire individuals with specific skills, education, and training, with the added benefit that they are not subject to promotion and transfer out of the bureau.

Training and certification are of critical importance to the potential effectiveness of fire prevention programs. Both are sometimes mandated for fire prevention personnel to be authorized to perform their official duties but are always desirable. Certification demonstrates initiative and technical competence on the part of the employee and organization.

## REVIEW QUESTIONS

1. List two organizations that provide accreditation for the certification of fire prevention bureau personnel.
2. The NFPA maintains standards for the professional qualifications for which four traditional fire prevention bureau positions?

3. What is the term used by NFPA in its standards for professional qualifications for job-specific skills used in the standards?
4. List two advantages of staffing fire prevention bureaus with uniformed firefighters.
5. List two advantages of staffing fire prevention bureaus with civilian personnel.
6. List two advantages of staffing fire prevention bureaus with a combination of both uniformed and civilian personnel.

## DISCUSSION QUESTIONS

You report for duty as the new fire marshal and find two telephone messages on your desk. The first one is from the city's director of planning and budget, requesting your attendance at a meeting to discuss the implementation of his plan to reduce costs by eliminating all the uniformed firefighter positions in the fire prevention bureau and replacing them with civilians whose inspector positions have been eliminated in other agencies as part of his "Reinventing Government" initiative.

The second message is from the new fire chief, requesting you develop a plan to staff the bureau with light-duty firefighters who will work in the bureau for up to a year and then return to full duty or retire on disability. The chief believes this is a good method of discouraging malingerers within the department.

1. Which call will you return first?
2. Develop a list of talking points for use in each meeting with your concerns for each proposed plan.

## CHAPTER PROJECT

Locate job announcements for fire prevention positions within government agencies and (or) private industry. Compare the minimum requirements or KSAs to the fire science curriculum and develop recommendations for students seeking employment in the fire prevention field.

## ADDITIONAL RESOURCES

In-depth information on many of the subjects discussed in this chapter can be found in the following texts and publications and at these Web sites.

Dennis Compton, et al., *Managing Fire and Rescue Services* (Washington, DC: International City/County Managers Association, 2002).

Steven T. Edwards, *Fire Service Personnel Management* (Upper Saddle River, NJ: Prentice Hall, 2004).

International Fire Service Accreditation Congress at www.ifsac.org.

National Board on Fire Service Professional Qualifications at www.theproboard.org.

## NOTES

1. *Standard for Professional Qualifications for Fire Inspector and Plan Examiner* (Quincy, MA: National Fire Protection Association, 2009), page 1.
2. Ibid., Appendix C, page 1.
3. Ibid., Section 4.3.2
4. Ibid., Section 5.3.1
5. Ibid., Section 6.3.1
6. Henry Campbell Black, et al., *Black's Law Dictionary* (St. Paul: West Publishing, 1991), page 401.

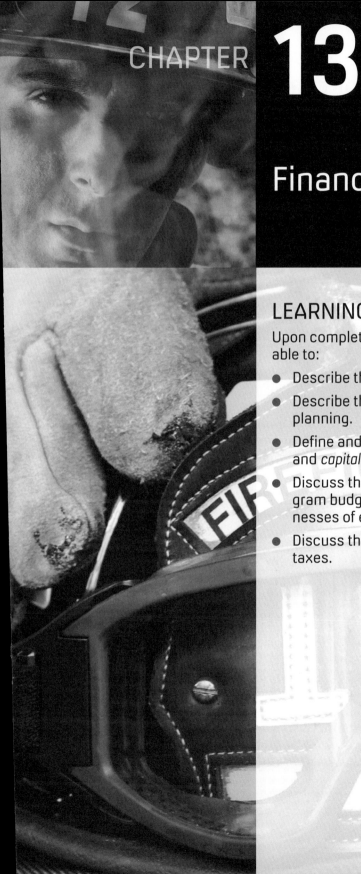

# 13

# Financial Management

## LEARNING OBJECTIVES

Upon completion of this chapter, you should be able to:

- Describe the purpose of public budgeting.
- Describe the use of a budget as a tool for planning.
- Define and describe the terms *operating budget* and *capital budget* or *capital improvement budget*.
- Discuss the use of line-item budgets and program budgets and the strengths and weaknesses of each.
- Discuss the difference between user fees and taxes.

# FINANCIAL MANAGEMENT OF FIRE PREVENTION PROGRAMS

Money management is one of the aspects of fire department and fire prevention bureau operations that rarely shows up on the radar unless or until you find yourself in the unenviable position of having to do it. Most of us tend to relegate public service financial management to a special category of knowledge and skills—important ones we do not need. Within that category are others such as sewage treatment, electrical generation, sausage manufacturing, and nuclear engineering. It is obviously important, we probably cannot live without it, somebody has to do it, and I am glad it is not me.

**NOTE** Most of us tend to relegate public service financial management to a special category of knowledge and skills—important ones we do not need.

The fault with this attitude is that money considerations must be a part of every management decision that is made. You cannot run the fire department or the fire prevention bureau without money. You cannot develop and manage a program that provides public services without considering value any more than you can go to the grocery store without looking at prices. There is an extra, extremely important consideration for public sector managers—it is not your money. We all have the right to squander our own money in any manner we choose but not those that belong to the public.

Public employees are in a position of trust. Public employees have a moral and legal obligation to give the citizens whose taxes fund the government their money's worth. They deserve the most bang for their buck that can be delivered. History is rife with instances of financial mismanagement by public officials. Some cases involve criminal conduct; others are merely instances of inept public officials. Laws and regulations regarding government financial management are the fruits of the mismanagement and corruption of the past.

**NOTE** Public employees have a moral and legal obligation to give the citizens whose taxes fund the government their money's worth.

## THE BUDGET PROCESS

**Budget** is a word that most firefighters do not want to hear. At home, it is a thing that reminds you of what you cannot have. At work, it reminds you that there are much worse jobs than anything you might have complained about in the fire department. Not many subjects will evoke total silence at a fire station kitchen table. A request for volunteers to assist with budget preparation is greeted with as much optimism as a firing squad. It should not be that way. Without effective budget preparation and disciplined fiscal management, government agencies cannot deliver adequate services to the public (Figure 13-1). That goes for fire prevention bureaus, fire departments, and every other government agency.

**budget** government financial plan that includes spending and income, often divided by categories for a given time, normally 1 year.

**FIGURE 13-1**
Funding for
this firefighter's
protective clothing,
equipment, and
salary all came from
his department's
budget. (Courtesy of
Duane Perry.)

The word *budget* comes from the Latin word *bulga*, which means "leather bag." Think of it as the bag of money used to fund a government program. The family budget is designed to keep spending in line with income and to ensure that the necessities are taken care of first. Government budgets are slightly different in that they are required by law, are used for planning purposes, and are actually used to determine how much money will be needed (through taxes, licenses, and fees) to fund the government. Imagine approaching your boss with: "My operating expenses [groceries and utilities] and capital expenses [house and car] have increased. I'll need a five percent increase in my paycheck effective July." It sounds ridiculous, but that is the way local governments set their real-estate tax rates and fee schedules.

State laws require local governments to balance their budgets. Deficit spending is not permitted. Managing an existing fire prevention program, attempting to expand a program, or starting from scratch requires skill in the budget process. Overspending, failing to recognize potential shortages during the fiscal year, and failure to include foreseeable expenses are signs of poor management and do not go unnoticed by elected officials and the public. It gets worse, however. There are laws regarding the spending of public money. What some may consider merely stretching the rules might well be illegal. Most of us worry about our credit report, at least when we apply for a loan. Governments also have credit reports, called their bond rating.

**NOTE** State laws require local governments to balance their budgets.

**NOTE** Governments also have credit reports, called their bond rating.

## Financial and Accounting Standards

In 1995, Washington, DC's bond rating was reduced to junk bond status.[1] Governments borrow money by selling bonds. The interest they must pay is based on their credit worthiness, or bond rating. To understand the impact of the reduction of Washington, DC's bond rating, imagine that you wanted to buy a house but could not qualify for the loan, so you had to buy it with your credit card. Washington, DC, was forced to approach the federal government for a loan in 1995 because a bond issue was not possible. The federal government loaned the district money, with strings attached. Congress took control of the city's finances by appointing a control board. A chief financial officer (CFO) was appointed by the control board to manage the city's finances. The city successfully turned things around, the CFO was later elected mayor, and by 2003, Washington, DC's bond rating had risen to A-.[2] By 2009, it was AAA, the highest rating by Standard and Poor's.

How are bond ratings determined? Governments are audited to protect against waste and fraud and as a means to measure the effectiveness of management. The Governmental Accounting Standards Board (GASB) was created in 1984 to establish standards for accounting by state and local governments. GASB standards for financial reporting are designed to aid auditors, financial institutions, and the public in assessing the financial accountability of governments. By using accounting standards, corruption is more difficult to conceal. Bond rating services, such as Moody's, Standard and Poor's, and Fitch Ratings Service, use the reports in their assessment of the government's financial management to determine the rating.

**NOTE** By using accounting standards, corruption is more difficult to conceal.

## Budget Basics

What do budgets and bond ratings have to do with the prevention of hostile fires? Financial management is a key part of the overall management of any government program. Well-run governments are sticklers when it comes to budget preparation. Their bond rating depends on it, and there are laws that require it. Funding for salaries, operating expenses, and equipment is secured through the budget process. The purpose of a budget is to:

- Supply information to elected officials in order for them to determine what programs and services will be funded and to what extent
- Supply information on government programs to the public, facilitating public participation in the process
- Establish a formal policy for spending public funds
- Assist managers by categorizing and tracking expenditures

The basic process of public budgeting is the same for all levels of government. The chief operating executive prepares a budget for the upcoming year. Included in the budget are detailed financial projections for government services, service levels, and the income required to provide the services. This **recommended budget** is then submitted to the elected governing body and is made available to the public. The governing body holds public hearings and receives public comment. The governing body has the option of modifying the recommended budget and voting to approve it. This approved package is the **adopted budget** for the **fiscal year** and includes any changes in taxes, licenses, or fees required to fund the programs.

**recommended budget** proposed budget submitted to the legislative body by the executive for action; when approved after public hearings, it becomes the approved budget.

**adopted budget** finalized budget that has been formally approved by the governing body of the jurisdiction.

**fiscal year** budget cycle year, usually July 1 through June 30; the federal fiscal year is October 1 through September 30.

The adopted budget for the jurisdiction typically consists of two sections, the **operating budget** and the capital budget or **capital improvement budget**. The operating budget includes funding for the day-to-day operating expenses of running the government. Personnel costs for salaries and fringe benefits, fuel, expendable supplies and equipment, rent, utilities, and interest payments are among the items included, generally divided into categories and called line items or funds (Figure 13-2). Transfers within the government or to other governments are also included in the operating budget. Local governments often transfer funds to school boards and public utilities or parks authorities. The federal budget includes transfers to the state governments. State governments transfer funds to counties and cities.

**operating budget** government financial plan that includes funding for the day-to-day operations of government, including salaries, utilities, rent, fuel, and transfer payments.

**capital improvement budget** long-range government financial plan typically used to fund construction projects, road construction, and land acquisition.

**FIGURE 13-2**
Funding for fire station construction is included in the capital or the capital improvement budget.

The capital or capital improvement budget is a long-range budget plan used to guide the financing and construction of capital projects. The capital budget includes funding for land acquisition, the development of parks, and the construction of roads. Construction and renovation projects for government buildings such as schools, libraries, and fire stations are included in the capital budget. Capital projects such as fire station construction are primarily funded with bonds. Jurisdictions borrow the money by issuing bonds. Bonds are actually promises to pay it back. The interest rate on the bonds is determined by the jurisdiction's bond rating.

**NOTE** Capital projects such as fire station construction are primarily funded with bonds.

Bonds issued by the U.S. government are the best there is. They are backed by the full faith and credit of the United States. If the Treasury Department has trouble paying, it calls the Bureau of Engraving and Printing and has it print more 50s and 100s. States, cities, and counties are not that fortunate, so they must rely on their excellent bond rating, the indicator of their future ability to pay off loans.

Bonds normally require voter approval in the form of a bond referendum. Voters are asked to cast a ballot in favor of or in opposition to the long-term debt. Some strategies are used by government managers to improve the chances for approval. Public perception of the agency has considerable impact on the process. Good customer relations usually translates into bond approval on election day. Bond referendums often include several projects and are usually included on the ballot with the election of officials. Projects on the ballot during contentious elections or with projects that evoke strong public opposition are often defeated, something like guilt by association. The fact that certain candidates are on the ballot brings out a large number of voters who vote no to everything they can.

**NOTE** Good customer relations usually translates into bond approval on election day.

## The Budget Tool

Unfortunately, many managers believe that once the budget is adopted, they are off the hook. The budget goes into a desk drawer or sits on the bookcase and collects dust. However, a budget is a tool that should be in constant use. Quarterly reviews that track spending should be carefully studied. Potential funding shortfalls in categories such as overtime pay must be identified and brought to the attention of appropriate authorities before the crises. A chief of fire prevention cannot control the fact that the caseload of investigators rose dramatically because of a serial arsonist. The chief can present the fact that at the present rate, the overtime budget will be exhausted at midyear. By tracking expenditures, there is ample time to request and justify additional funding.

**NOTE** A budget is a tool that should be in constant use.

Experienced managers also learn that trends are reoccurring. In most governments, shortfalls occur in certain categories every year. For various reasons, some budget categories are intentionally underfunded. Public safety overtime pay is one of those categories. Budget directors often theorize that public safety agencies will spend whatever they get, and then some. Underfunding is a tactic used to reduce that tendency.

Toward the end of the third quarter of the fiscal year, agency heads identify looming shortfalls and notify the budget director. The initial method of funding the shortage is by transferring funds from nonessential categories such as training and education. Seasoned managers schedule training activities and procure supplies and equipment in the first half of the fiscal year. They have a wish list for training and equipment on hand for use at the end of the year in case other branches or agencies have surplus funding.

**NOTE** Seasoned managers schedule training activities and procure supplies and equipment in the first half of the fiscal year.

## Budget Calendar

The federal fiscal year runs October 1 through September 30. Most other levels of government use July 1 through June 30. Budget preparation typically begins in late summer, when county and city managers direct their agency heads to begin work on the budget for the following fiscal year. Executives typically provide guidelines that limit spending increases and include inflation factors and costs estimates from purchasing schedules.

New programs and changes in service levels that require additional staffing typically must be justified in the earliest stages of the budget cycle. In practical terms, the chief fire marshal who wants to establish a new program needs to have the fire chief's approval almost a year before the program's proposed start date. If it takes 6 months to develop statistical information to get the fire chief's approval, the timeline stretches 18 months. Many managers lack the foresight and the determination required to bring new programs to life. The process can be grueling, particularly in austere times. Competition with other agencies or even within the fire department for funding new programs is a hurdle few firefighters ever envisioned when they started their careers.

## The Budget as an Evaluation Tool

Several types of budgets have evolved over the years. There are different ways of developing a long-range planning tool with citizen input that converts long-range plans into viable services and programs. Each attempt has been aimed at improving the process and overcoming a basic problem that elected officials have grappled with in every democracy: how to determine the services that people really need, how much to spend, and the quality of services delivered. In the private sector, it is pretty simple. Restaurants with bad food and surly waiters do not stay in business very long. Word gets out, and people go elsewhere. It is survival of the fittest in the free market.

Government lacks that built-in mechanism for evaluation, so methods of measuring performance have been included in the budget process with different levels of intensity. In 1993, Congress enacted the *Government Performance and Results Act.* In the Senate's report that accompanied the legislation, members of the Committee on Government Affairs cited the "inconsistency between the public's desire for a wide range of government services," and the "public's disdain for government and objection to paying higher taxes."[3] The Committee went on to say, "The Committee shares the public's frustration with waste, inefficiency, and ineffectiveness" of government programs.[4]

The act required federal agencies to develop methods of measuring performance in delivery of services. The idea was not new, however. Management indicators have long been a part of the budget process in many jurisdictions. Typically, the costs of delivering a service or staff hours per unit were compared from year to year. The object was to deliver more with less. Quality was often another matter, but it is now included in many government systems that measure performance.

## Budget Formats

Books on government financial management usually describe eight types of government budget formats, some of which are quite similar. The budget systems of many jurisdictions actually are hybrids—they contain aspects of several formats. The five most radically different budget types are the line-item budget, lump-sum budget, program budget, performance budget, and zero-based budget.

### Line-Item Budget

The **line-item** budget is the most common budget format in use by governments,[5] and many jurisdictions use some form of the line-item budget. Each line item is a fund or subfund. A typical fire prevention bureau budget has a fund category for books and printed material, with a subfund or line item for codebooks. There may also be line items for hand tools and equipment, educational material for public distribution, uniforms, and fuel for the bureau's vehicles. There should be a line item for each category of operational expenses.

| line item | fund category within a budget. |
|-----------|-------------------------------|

The line-item budget provides strong control over spending because funds cannot be shifted between categories without approval. It is easy to track expenditures because each expenditure is identified by line item. Agency managers justify spending by category. The simplicity of the system and the strong fiscal control tend to promote unnecessary spending. Money that remains unspent at the end of the fiscal year is returned to the general fund. Not spending all the money within a category invariably leads to a dreaded comment from the budget analyst during the next budget cycle: "You did not need the money this year; why can't you get along without it next year?"

**NOTE** The line-item budget provides strong control over spending because funds cannot be shifted between categories without approval.

## Lump-Sum Budget

In *Managing Fire Services*, Douglas Ayers and Leonard Marks note that before the governmental reform movement of the early 1900s, fire chiefs and other municipal agency directors were allocated funds in one lump sum.[6] The fire chief, normally a political appointee, spent the money as he or she saw fit to run the fire department. It is not hard to imagine some of the abuses that took place, and it is easy to see that the problems created by the lump-sum budget led to the development of the line-item budget.

## Program Budget

The program budget was developed to streamline the cumbersome line-item budget process. Funding is allocated by major program area or service. Line items are eliminated or greatly reduced and replaced with general categories such as personnel costs, operating equipment, and capital construction costs. The object was to place less emphasis on the minute details of the line-item budget and more emphasis on goal setting and service delivery.

## Performance Budget

The performance budget was the first to include specific standards of performance for government agencies. Standards are developed for each functional area, and performance is compared with that of previous years. The object is to improve each year. These categories are often referred to as *management indicators* or *performance indicators*. Care must be taken to develop performance indicator categories that accurately reflect the agency's mission and that the agency can control. Mistakes can literally derail the mission of the organization.

Fire investigations units are charged with determining fire cause and origin and follow-up investigation. A finding of criminal activity leads to a referral to the criminal justice system. The most that can be expected from the unit is an accurate determination of fire cause, rapid follow-up investigation, and arrest of the suspects in instances of criminal activity. From that point on, the case is outside the control of the fire investigation unit. The decision to pursue the court case, performance during the trial, and ultimate outcome regarding conviction belong to others. The prosecuting attorney, judge, and jury will make the determinations. Rate of conviction is not an accurate indicator of performance because it measures events outside the realm of the unit's control.

The crime of arson is frequently committed in connection with other crimes. Prosecuting attorneys may determine that the best course is to accept a plea agreement in which the arson charge is dropped in return for guilty pleas to other charges. Imagine the scenario if rate of conviction is used as an indicator: Good investigative work on the fire scene uncovers a burglary ring that uses fire to cover its tracks. Fire investigators apprehend the suspects and break the case. During the trial, the suspects realize that an airtight case has been presented. Before jury deliberation, they agree to plead guilty to multiple counts of burglary and associated crimes, with lengthy prison terms, if the arson charges are dropped. The unit has performed flawlessly and successfully performed its mission, but a budget analyst would disagree. Without an arson conviction, the unit failed to perform.

Functions that are controlled by the organization but are not central or core elements of the organization's mission must also be avoided. When a function is adopted as a budget performance indicator, it instantly becomes a central part of the organization's mission. Managers shift resources from other areas to bolster the numbers; the agency's financial future depends on it.

## Zero-Based Budget

In developing a zero-based budget, all government programs are rated in order of priority. Funding is then allocated based on the value of the service provided. The system provides a complete rejustification of each program every year. Determining the relative value or priority of programs can lead to considerable infighting within the government, with the inherent risk of damage to cooperative relationships between agencies.

In the private sector, a corporation is better able to prioritize functions. Which functions generate more profit or support the functions that do? Making that determination for government services is much harder. Is it more important to provide fire or police protection? Is water and sewer service more important than electricity? In addition to the difficulty in prioritizing services, zero-based budgeting is the most labor intensive, requiring a complete rejustification of every program every year.

# Fire Prevention Bureau Budgets

Fire prevention bureaus have duties and responsibilities that differ considerably from the rest of the fire department. The budget for the fire inspections branch more closely resembles that of the building department. Personal protective equipment, weapons and ammunition, and evidence collection kits used by fire investigators result in a budget that looks more like a police agency's than a fire department's.

Fire department budget analysts and some senior fire department officials often question the need for funding in unfamiliar categories. Regardless of the frustration such questions will cause, it is their job to ensure that spending is necessary.

**NOTE** Fire department budget analysts and some senior fire department officials often question the need for funding in unfamiliar categories.

I don't remember any gunfights. Why buy all this ammunition?
*Investigators must requalify with their weapons every year.*

We purchased codebooks for every member of the bureau three years ago. Did they wear them out?
*The most recent edition of the code is scheduled for adoption next year. The old books will no longer be the adopted code.*

## Replacement Schedules

Developing replacement schedules for tools and equipment is good policy. It is much easier to justify the replacement of one-third of the bureaus cameras every 3 years than to request total replacement every 3 years. One problem is that you have not asked for cameras for the past 2 years. Did the mission change? Why start taking pictures now? With the replacement schedule, the hard work of justifying the cameras is done once. The same justification will be resubmitted every year.

**NOTE** Developing replacement schedules for tools and equipment is good policy.

Developing close working relationships with other agencies also eases the budget burden. The police department's justification, pricing schedule, and technical specifications for their weapons purchases are only a phone call away. The building department's training budget and technical reference and code budget can also prove to be invaluable. Good working relationships on the street should translate into cooperative relationships during the budget process. The police and fire department double-team may be more than the budget analyst can handle.

**NOTE** Good working relationships on the street should translate into cooperative relationships during the budget process.

## COST RECOVERY AND FEE STRUCTURES

The use of user fees to pay for government services has gained widespread acceptance throughout all levels of government. In many cases, government agencies have the solid backing of the business community, provided it receives prompt, efficient service. For business, time is usually money. Delays in inspections or approvals translate into cost overruns. Real estate developers typically finance the entire cost of construction, including inspection fees. Delays in plan review, permit processing, or inspection can literally freeze an entire project in its tracks. Developers and business financial managers generally believe that fees are the cost of doing business. They want to know how much and how fast ahead of time.

Local governments often require enabling legislation from the state legislature to establish user fee systems. A key point that is sometimes brushed aside by agency managers is that a **user fee** is not a tax. User fees are payments for service. Legislative bodies such as state legislatures and city and county councils cannot and do not delegate their legal right to levy taxes. User fees must be based on some logical system such as cost to perform the service.

**user fee** fee for service charged by a government organization for providing a specific service.

**NOTE** User fees are payments for service.

# The Fee Adoption Process

The implementation of a fee-for-service system typically requires a legal enablement at the state level and adoption by local elected officials. The process of local adoption is time consuming and requires significant staff resources. Proper planning and good communication with all those affected will ensure success. Most fire departments do not have in-depth experience with billing, collection, and accounting. Answers to the who, where, and how questions that will invariably arise during hearings and public comment must be developed and evaluated against all possibilities. It only takes one unanswered question during adoption hearings to derail the process.

The strategy guaranteed to deliver unsatisfactory results is the establishment of a fee system without the input of the business community. Attempts to quietly move the program through the political process will most likely be met by an all-out assault from the business community. Elected officials, wary of being stuck between their constituents and agency officials, may determine that they have been used. Future attempts at garnering their support might fall on deaf ears.

**NOTE** The strategy guaranteed to deliver unsatisfactory results is the establishment of a fee system without the input of the business community.

The business community has and will certainly exercise the right to address the elected body regarding the establishment of fees. The best course for the fire prevention bureau is to keep them involved from the beginning. With business support, the adoption of a fee schedule is almost guaranteed. Adoption may occur without business support but with the acknowledgement that they were given opportunity for input, provided elected officials agree on the need. Accusations of secrecy or that the business community was blindsided will guarantee defeat and prove embarrassing. When government officials use such tactics, the public tends to remember.

**NOTE** With business support, the adoption of a fee schedule is almost guaranteed.

# Determining the Fee Structure

What is a fair method of determining user fees? Fire prevention bureaus that develop inspections fees might develop a cost-per-hour figure and base the fee schedule on the average time to perform an inspection. The cost per hour is based on budget data. Salary, fringe benefits, transportation, and a share of the bureau's operating expenses are used to develop an average cost per hour or per inspection. Inspection or permit fees are based on this figure. By using data that are available to the general public through the budget, the system is aboveboard.

The next, perhaps more important, element to consider is the quality of service provided. Business and the public deserve a quality product in return for the fee. Because the user fee is not a tax, the public has every right to demand a quality service. Poor-quality service opens the door to privatization.

## Types of Fees

User fees can be developed for most, but not all, functions performed by fire prevention bureaus. The state enabling legislation is usually worded to the effect "local governments may defray the cost of inspection and enforcement of the code through the establishment of fees." User fees for the fire investigations functions probably do not fit into that category. You probably would not have much luck collecting a fee from a guy arrested for torching his car anyway. Typical fees collected by fire prevention bureaus include the following.

### Permit Fees

The primary purpose of the model fire code permit systems is as a means to require inspection by the fire official. Permits are required for occupancies and processes with a high probability of fire incidence, large potential for injury or death because of high occupant loads, or high hazard to the public because of large quantities of regulated materials. Permits are also required to install, modify, extend, or remove fire protection systems. Additionally, permits serve to:

- Notify first responders of the presence of dangerous materials and processes.
- Provide leverage for compliance through the threat of revocation.
- Provide leverage for access through the threat of revocation.

Typically, permit fees are based on the average time required to inspect the project involving the permit. Many facilities require multiple permits.

### Inspection Fees

Many jurisdictions require inspections that do not involve the permit system. Inspections of new occupancies or at change of tenancy are common. Inspections involving licensure by other agencies, including institutional occupancies and daycare centers, are commonly performed by fire prevention bureaus. Many are not required by the fire code but are mandated within the regulations of the licensing agencies. Fees are typically assessed based on the time required to perform the inspection and prepare the inspection report.

### Plan Review Fees

Within many jurisdictions, fire protection system plans, site plans, and building plans are reviewed by fire prevention bureaus. Fee schedules are typically based on plan type, square footage, number of fire protection devices, or reviewer's time. Fees for additional review when plans are rejected must be developed to ensure an equitable treatment of all those involved. Flat fees for additional reviews tend to reward a few design professionals and contractors for shoddy, incomplete work at the expense of the conscientious members of the profession. If additional reviews cost less than the initial fee, why get it right the first time? Multiple reviews also increase reviewer workload, adding days or weeks to already tight schedules.

## Acceptance Testing and Retesting Fees

Required acceptance tests and retests of fire protection systems must be conducted in the presence of fire prevention bureau representatives in some jurisdictions to ensure that the systems are properly installed and operate as designed. Owners of large complexes and some developers have found that fees associated with fire protection system tests are the best insurance against poor workmanship and improper installation by contractors. Fees are generally per hour or per device. Fees for retests or reinspection must be established with the same care as plan review fees. Low, fixed retest fees are an invitation for installing contractors not to pretest their work.

**NOTE** Owners of large complexes and some developers have found that fees associated with fire protection system tests are the best insurance against poor workmanship and improper installation by contractors.

## Implementation

The time for establishing a tracking and accounting system with reports and safeguards is not after the program is adopted. State laws and regulations for government financial management must be strictly followed. Most likely, there is already a system in place within the jurisdiction. Assistance from city or county finance officials should be sought early on in the process to minimize duplication of preexisting systems or the development of conflicts within systems.

Existing software programs, accounting procedures, and security measures may be easily adapted for use, ensuring integration with the jurisdiction's system of financial management. Good security and formal procedures help keep honest people honest.

**NOTE** Good security and formal procedures help keep honest people honest.

## SUMMARY

Public employees are in positions of trust. Public employees have a moral and legal obligation to give good value and fair accounting to the citizens whose taxes fund government programs. History is rife with instances of financial mismanagement by public officials. Laws and regulations regarding government financial management are the fruits of the mismanagement and corruption of the past.

The purpose of a budget is to:

- Supply information to elected officials in order for them to determine what programs and services will be funded and to what extent
- Supply information on government programs to the public, facilitating public participation in the process
- Establish a formal policy for spending public funds
- Assist managers by categorizing and tracking expenditures

Government budgets are generally divided into the operating budget and the capital or capital improvement budget. The operating budget contains funding for the day-to-day operations of the government. The capital or capital improvement budget is a long-range planning tool used to fund projects such as land for road, park, or facility construction.

The establishment of user fee systems is a method of providing services that protect taxpayers by only charging those who receive the service. Inspection, plan review, and fire protection system acceptance testing are fire prevention bureau functions that are funded through user fees in some jurisdictions. Enabling legislation on the state level and adoption of a fee schedule on a local level are required before implementation. State laws and regulations for accounting, tracking, reporting, and security must be in full compliance.

## REVIEW QUESTIONS

1. List the two broad types of budget.
2. List five budget formats.
3. Define the term *line item*.
4. Define the terms *performance indicator* and *management indicator*.
5. Describe a method of determining reasonable inspection fees.
6. Define the term *fiscal year* and explain its significance in the budget process.

## DISCUSSION QUESTIONS

1. Because of the untimely death of the administrative captain, you have been tasked with budget preparation for the fire prevention bureau. Develop a list of agencies and officials within a local government that could be consulted for assistance, guidance, and information.
2. You have been tasked with developing a user fee system to defray the cost of providing inspection and plan review services. During a meeting with the chief, you recommended meeting with the business community and seeking their input. The chief agreed and told you to "handle it." Develop a short description of the fee system you propose. Use the Internet or phone directory to develop a list of local organizations you will contact.

## CHAPTER PROJECT

Review the budget for the fire prevention bureau within your jurisdiction. Develop strategies to reduce costs to the taxpayer while increasing the level of service. For any proposed increases in fees, develop strategies to garner the support of the affected constituencies.

## ADDITIONAL RESOURCES

In-depth information on many of the subjects discussed in this chapter can be found in the following texts and publications and at these Web sites.

Dennis Compton, et al., *Managing Fire and Rescue Services* (Washington, DC: International City/County Managers Association, 2002).

Government Accounting Standards Board (GASB) at www.gasb.org.

Government Finance Officers Association (GFOA) of www.gfoa.org.

National Association of Counties (NACO) at www.naco.org.

National Association of State Treasurers (NAST) at www.nast.net.

## NOTES

1. David Nakamura, "District Gets A- Rating," *The Washington Post*, June 24, 2003, page B2.
2. Ibid.
3. Committee on Governmental Affairs, United States Senate, *Senate Report 103-58, Government Performance and Results Act of 1993 Report*, page 2.
4. Ibid.
5. Dennis Compton, et al., *Managing Fire and Rescue Services* (Washington, DC: International City/County Managers Association, 2002), page 197.
6. Ibid., page 196.

Public Law 90-259, enacted by the 90th Congress and signed by President Richard M. Nixon, was a significant milestone in the history of fire prevention and fire protection in the United States.

## PUBLIC LAW 90-259

(90th Congress, S. 1124, Mar. 1, 1968)
AN ACT

To amend the Organic Act of the National Bureau of Standards to authorize a fire research and safety program, and for other purposes.

*Be it enacted by the Senate and House of Representatives of the United States of America in Congress assembled,* That this Act may be cited as the "Fire Research and Safety Act of 1968."

## TITLE I—FIRE RESEARCH AND SAFETY PROGRAM

### Declaration of Policy

SEC. 101. The Congress finds that a comprehensive fire research and safety program is needed in this country to provide more effective measures of protection against the hazards of death, injury, and damage to property. The Congress finds that it is desirable and necessary for the Federal Government, in carrying out the provisions of this title, to cooperate with and assist public and private agencies. The Congress declares that the purpose of this title is to amend the Act of March 3, 1901, as amended, to provide a national fire research and safety program including the gathering of comprehensive fire data; a comprehensive fire research program; fire safety education and training programs; and demonstrations of new approaches and improvements in fire prevention and control, and reduction of death, personal injury, and property damage. Additionally, it is the sense of

Congress that the Secretary should establish a fire research and safety center for administering this title and carrying out its purposes, including appropriate fire safety liaison and coordination.

## Authorization of Program

SEC. 102. The Act entitled "An Act to establish the National Bureau of Standards", approved March 3, 1901, as amended (15 U.S.C. 271-278e), is further amended by adding the following sections:

"SEC. 16. The Secretary of Commerce (hereinafter referred to as the 'Secretary') is authorized to—

"(a)  Conduct directly or through contracts or grants-

"(1)  investigations of fires to determine their causes, frequency of occurrence, severity, and other pertinent factors;

"(2)  research into the causes and nature of fires, and the development of improved methods and techniques for fire prevention, fire control, and reduction of death, personal injury, and property damage;

"(3)  educational programs to-

"(A)  inform the public of fire hazards and fire safety techniques, and

"(B)  encourage avoidance of such hazards and use of such techniques;

"(4)  fire information reference services, including the collection, analysis, and dissemination of data, research results, and other information, derived from this program or from other sources and related to fire protection, fire control, and reduction of death, personal injury, and property damage;

"(5)  educational and training programs to improve, among other things-

"(A)  the efficiency, operation, and organization of fire services, and

"(B)  the capability of controlling unusual fire-related hazards and fire disasters; and

"(6)  projects demonstrating-

"(A)  improved or experimental programs of fire prevention, fire control, and reduction of death, personal injury, and property damage,

"(B)  application of fire safety principles in construction, or

"(C)  improvement of the efficiency, operation, or organization of the fire services.

"(b)  Support by contracts or grants the development, for use by educational and other nonprofit institutions, of-

"(1)  fire safety and fire protection engineering or science curriculums; and

"(2)  fire safety courses, seminars, or other instructional materials and aids for the above curriculums or other appropriate curriculums or courses of instruction.

"**SEC. 17.** With respect to the functions authorized by section 16 of this Act-

"(a)   Grants may be made only to States and local governments, other non-Federal public agencies, and nonprofit institutions. Such a grant may be up to 100 per centum of the total cost of the project for which such grant is made. The Secretary shall require, whenever feasible, as a condition of approval of a grant, that the recipient contribute money, facilities, or services to carry out the purpose for which the grant is sought. For, the purposes of this section, 'State' means any State of the United States, the District of Columbia, the Commonwealth of Puerto Rico, the Virgin Islands, Guam, the Canal Zone, American Samoa, and the Trust Territory of the Pacific Islands; and 'public agencies' includes combinations or groups of States or local governments.

"(b)   The Secretary may arrange with and reimburse the heads of other Federal departments and agencies for the performance of any such functions, and, as necessary or appropriate, delegate any of his powers under this section or section 16 of this Act with respect to any part thereof, and authorize the redelegation of such powers.

"(c)   The Secretary may perform such functions without regard to section 3648 of the Revised Statutes (31 U.S.C. 529).

"(d)   The Secretary is authorized to request any Federal department or agency to supply such statistics, data, program reports, and other materials as he deems necessary to carry out such functions. Each such department or agency is authorized to cooperate with the Secretary and, to the extent permitted by law, to furnish such materials to the Secretary. The Secretary and the heads of other departments and agencies engaged in administering programs related to fire safety shall, to the maximum extent practicable, cooperate and consult in order to insure fully coordinated efforts.

"(e)   The Secretary is authorized to establish such policies, standards, criteria, and procedures and to prescribe such rules and regulations as he may deem necessary or appropriate to the administration of such functions or this section, including rules and regulations which-

"(1)   provide that a grantee will from time to time, but not less often than annually, submit a report evaluating accomplishments of activities funded under section 16, and

"(2)   provide for fiscal control, sound accounting procedures, and periodic reports to the Secretary regarding the application of funds paid under section 16."

## Noninterference with Existing Federal Programs

**SEC.** 103. Nothing contained in this title shall be deemed to repeal, supersede, or diminish existing authority or responsibility of any agency or instrumentality of the Federal Government.

## Authorization of appropriations

**SEC.** 104. There are authorized to be appropriated, for the purposes of this Act, $5,000,000 for the period ending June 30, 1970.

## TITLE II—NATIONAL COMMISSION ON FIRE PREVENTION AND CONTROL

## Findings and Purpose

**SEC.** 201. The Congress finds and declares that the growing problem of the loss of life and property from fire is a matter of grave national concern; that this problem is particularly acute in the Nation's urban and suburban areas where an increasing proportion of the population resides but it is also of national concern in smaller communities and rural areas; that as population concentrates, the means for controlling and preventing destructive fires has become progressively more complex and frequently beyond purely local capabilities; and that there is a clear and present need to explore and develop more effective fire control and fire prevention measures throughout the country in the light of existing and foreseeable conditions. It is the purpose of this title to establish a commission to undertake a thorough study and investigation of this problem with a view to the formulation of recommendations whereby the Nation can reduce the destruction of life and property caused by fire in its cities, suburbs, communities, and elsewhere.

## Establishment of Commission

**SEC.** 202. (a) There is hereby established the National Commission on Fire Prevention and Control (hereinafter referred to as the "Commission") which shall be composed of twenty members as follows: the Secretary of Commerce, the Secretary of Housing and Urban Development, and eighteen members appointed by the President. The individuals so appointed as members (1) shall be eminently well qualified by training or experience to carry out the functions of the Commission, and (2) shall be selected so as to provide representation of the views of individuals and organizations of all areas of the United States concerned with fire research, safety, control, or prevention, including representatives drawn from Federal, State, and local governments, industry, labor, universities, laboratories, trade associations, and other interested institutions or organizations. Not more than six members of the Commission shall be appointed from the Federal Government. The President shall designate the Chairman and Vice Chairman of the Commission.

(b)  The Commission shall have four advisory members composed of-

    (1)  two Members of the House of Representatives who shall not be members of the same political party and who shall be appointed by the Speaker of the House of Representatives, and

    (2)  two Members of the Senate who shall not be members of the same political party and who shall be appointed by the President of the Senate.

The advisory members of the Commission shall not participate, except in an advisory capacity, in the formulation of the findings and recommendations of the Commission.

(c) Any vacancy in the Commission or in its advisory membership shall not affect the powers of the Commission, but shall be filled in the same manner as the original appointment.

## Duties of the Commission

SEC. 203. (a) The Commission shall undertake a comprehensive study and investigation to determine practicable and effective measures for reducing the destructive effects of fire throughout the country in addition to the steps taken under sections 16 and 17 of the Act of March 3, 1901 (as added by title I of this Act). Such study and investigation shall include, without being limited to-

(1) a consideration of ways in which fires can be more effectively prevented through technological advances, construction techniques, and improved inspection procedures;

(2) an analysis of existing programs administered or supported by the departments and agencies of the Federal Government and of ways in which such programs could be strengthened so as to lessen the danger of destructive fires in Government-assisted housing and in the redevelopment of the Nation's cities and communities;

(3) an evaluation of existing fire suppression methods and of ways for improving the same, including procedures for recruiting and soliciting the necessary personnel;

(4) an evaluation of present and future needs (including long-term needs) of training and education for fire-service personnel;

(5) a consideration of the adequacy of current tire communication techniques and suggestions for the standardization and improvement of the apparatus and equipment used in controlling fires;

(6) an analysis of the administrative problems affecting the efficiency or capabilities of local fire departments or organizations: and

(7) an assessment of local, State, and Federal responsibilities in the development of practicable and effective solutions for reducing fire losses.

(b) In carrying out its duties under this section the Commission shall consider the results of the functions carried out by the Secretary of Commerce under sections 16 and 17 of the Act of March 3, 1901 (as added by title I of this Act), and consult regularly with the 'Secretary in order' to coordinate the work of the Commission and the functions carried out under such sections 16 and 17.

(c) The Commission shall submit to the President and to the Congress a report with respect to its findings and recommendations not later than two years after the Commission has been duly organized.

## Powers and Administrative Provisions

**SEC.** 204. (a) The Commission or, on the authorization of the Commission, any subcommittee or member thereof, may, for the purpose of carrying out the provisions of this title, hold hearings, take testimony, and administer oaths or affirmations to witnesses appearing before the Commission or any subcommittee or member thereof.

(b) Each department, agency, and instrumentality of the executive branch of the Government, including an independent agency, is authorized to furnish to the Commission, upon request made by the Chairman or Vice Chairman, such information as the Commission deems necessary to carry out its functions under this title.

(c) Subject to such rules and regulations as may be adopted by the Commission, the Chairman, without regard to the provisions of title 5, United States Code, governing appointments in the competitive service, and without regard to the provisions of chapter 51 and subchapter III of chapter 53 of such title relating to classification and General Schedule pay rates, shall have the power-

(1) to appoint and fix the compensation of such staff personnel as he deems necessary, and

(2) to procure temporary and intermittent services to the same extent as is authorized by section 3109 of title 5, United States Code.

## Compensation of Members

**SEC.** 205. (a) Any member of the Commission, including a member appointed under section 202 (b) who as a Member of Congress or in the executive branch of the Government shall serve without compensation in addition to that received in his regular employment, but shall be entitled to reimbursement for travel, subsistence, and other necessary expenses incurred by him in connection with the performance of duties vested in the Commission.

(b) Members of the Commission, other than those referred to in subsection (a), shall receive compensation at the rate of $100 per day for each day they are engaged in the performance of their duties as members of the Commission and shall be entitled to reimbursement for travel, subsistence, and other necessary expenses incurred by them in the performance of their duties as members of the Commission.

## Expenses of the Commission

**SEC.** 206. There are authorized to be appropriated, out of any money in the Treasury not otherwise appropriated, such sums as may be necessary to carry out this title.

## Expiration of the Commission

SEC. 207. The Commission shall cease to exist thirty days after the submission of its report under section 203 (c)

Approved March 1, 1968.

### *Legislative history*

HOUSE REPORT No. 522 accompanying H.R. 11284 (Comm. on Science and Astronautics).

SENATE REPORT No. 502 (Comm. on Commerce).

CONGRESSIONAL RECORD: Vol. 113 (1967): Aug. 16, considered and passed Senate. Vol. 114 (1968): Feb. 8, considered and passed House, amended, in lieu of H.R. 11284. Feb. 16, Senate agreed to House amendment.

On May 4, 1973, the report of the United States Commission on Fire Prevention and Control transmitted the report of its findings pursuant to Public Law 90-259 (see Appendix A) to the president. The 177-page report, titled *America Burning*, set the stage for federal fire prevention and protection programs and galvanized state and local efforts. The 90 recommendations of the commission, published as Appendix IV of the report, are reproduced here. The entire text is available from the Government Printing Office or online from the U.S. Fire Administration in PDF format at www.usfa.fema.gov.

## CHAPTER 1

1. . . . . the Commission recommends that Congress establish a U.S. Fire Administration to provide a national focus for the Nation's fire problem and to promote a comprehensive program with adequate funding to reduce life and property loss from fire.

2. . . . * the Commission recommends that a national fire data system be established to provide a continuing review and analysis of the entire fire problem.

## CHAPTER 2

3. The Commission recommends that Congress enact legislation to make possible the attainment of 25 burn units and centers and 90 burn programs within the next 10 years.

4. The Commission recommends that Congress, in providing for new burn treatment facilities, make adequate provision for the training and continuing support of the specialists to staff these facilities. Provision should also be made for special training of those who provide emergency care for burn victims in general hospitals.

5. The Commission recommends that the National Institutes of Health greatly augment their sponsorship of research on burns and burn treatment.

6. The Commission recommends that the National Institutes of Health administer and support a systematic program of research concerning smoke inhalation injuries.

## CHAPTER 3

7. The Commission recommends that local governments make fire prevention at least equal to suppression in the planning of fire department priorities.

8. The Commission recommends that communities train and utilize women for fire service duties.

9. The Commission recommends that laws which hamper cooperative arrangements among local fire jurisdictions be changed to remove the restrictions.

10. The Commission recommends that every local fire jurisdiction prepare a master plan designed to meet the community's present and future needs in fire protection, to serve as a basis for program budgeting, and to identify and implement the optimum cost-benefit solutions in fire protection.

11. The Commission recommends that Federal grants for equipment and training be available only to those fire jurisdictions that operate from a federally approved master plan for fire protection.

12. The Commission recommends that the proposed U.S. Fire Administration act as a coordinator of studies of fire protection methods and assist local jurisdictions in adapting findings to their fire protection planning.

## CHAPTER 4

13. The Commission recommends that the proposed U.S. Fire Administration provide grants to local fire jurisdictions for developing master plans for fire protection. Further, the proposed U.S. Fire Administration should provide technical advice and qualified personnel to local fire jurisdictions to help them develop master plans.

## CHAPTER 5

14. The Commission recommends that the proposed U.S. Fire Administration sponsor research in the following areas: productivity measure of fire departments, lob analyses, firefighter injuries, and fire prevention efforts.

15. The Commission urges the Federal research agencies, such as the National Science Foundation and the National Bureau of Standards, to sponsor research appropriate to their respective missions within the areas of productivity of fire departments, causes of firefighter injuries, effectiveness of fire prevention efforts, and the skills required to perform various fire department functions.

16. The Commission recommends that the Nation's fire departments recognize advanced and specialized education and hire or promote persons with experience at levels commensurate with their skills.

17. The Commission recommends a program of Federal financial assistance to local fire services to upgrade their training.

18. In the administering of Federal funds for training or other assistance to local fire departments, the Commission recommends that eligibility be limited to those departments that have adopted an effective, affirmative action program related to the employment and promotion of members of minority groups.

19. The Commission recommends that fire departments, lacking emergency ambulance, paramedical and rescue services, consider providing them, especially if they are located in communities where these services are not adequately provided by other agencies.

# CHAPTER 6

20. The Commission recommends the establishment of a National Fire Academy to provide specialized training in areas important to the fire services and to assist State and local jurisdictions in their training programs.

21. The Commission recommends that the proposed National Fire Academy assume the role of developing, gathering, and disseminating, to State and local arson investigators, information on arson incidents and on advanced methods of arson investigations.

22. The Commission recommends that the National Fire Academy be organized as a division of the proposed US. Fire Administration, which would assume responsibility for deciding details of the Academy's structure and administration.

23. The Commission recommends that the full cost of operating the proposed National Fire Academy and subsidizing the attendance of fire service members be borne by the Federal Government.

# CHAPTER 7

24. The Commission urges the National Science Foundation, in its Experimental Research and Development Incentives Program, and the National Bureau of Standards, in its Experimental Technology Incentives Program, to give high priority to the needs of the fire services.

25. The Commission recommends that the proposed U.S. Fire Administration review current practices in terminology, symbols, and equipment descriptions, and seek to introduce standardization where it is lacking.

26. The Commission urges rapid implementation of a program to improve breathing apparatus systems and expansion of the program's scope where appropriate.

27. The Commission recommends that the proposed U.S. Fire Administration undertake a continuing study of equipment needs of the fire services, monitor research and development in progress, encourage needed research and development, disseminate results, and provide grants to fire departments for equipment procurement to stimulate innovation in equipment design.

28. The Commission urges the Joint Council of National Fire Service Organizations to sponsor a study to identify shortcomings of firefighting equipment and the kinds of research, development, or technology transfer that can overcome the deficiencies.

## CHAPTER 8

No recommendations.

## CHAPTER 9

29. The Commission recommends that research in the basic processes of ignition and combustion be strongly increased to provide a foundation for developing improved test methods.

30. This Commission recommends that the new Consumer Product Safety Commission give a high priority to the combustion hazards of materials in their end use.

31. ... the Commission recommends that the present fuel load study sponsored by the General Services Administration and conducted by the National Bureau of Standards be expanded to update the technical study of occupancy fire loads.

32. The Commission recommends that flammability standards for fabrics be given high priority by the Consumer Product Safety Commission,

33. The Commission recommends that all States adopt the Model State Fireworks Law of the National Fire Protection Association, thus prohibiting all fireworks except those for public displays.

34. The Commission recommends that the Department of Commerce be funded to provide grants for studies of combustion dynamics and the means of its control.

35. The Commission recommends that the National Bureau of Standards and the National Institutes of Health cooperatively devise and implement a set of research objectives designed to provide combustion standards for materials to protect human life.

## CHAPTER 10

36. The Commission urges the National Bureau of Standards to assess current progress in fire research and define the areas in need of additional investigation. Further, the Bureau should recommend a program for translating research results into a

systematic body of engineering principles and, ultimately, into guidelines useful to code writers and building designers.

37. The Commission recommends that the National Bureau of Standards, in cooperation with the National Fire Protection Association and other appropriate organizations, support research to develop guidelines for a systems approach to fire safety in all types of buildings.

38. The Commission recommends that, in all construction involving Federal money, awarding of those funds be contingent upon the approval of a fire safety systems analysis and a fire safety effectiveness statement.

39. This Commission urges the Consumer Product Safety Commission to give high priority to matches, cigarettes, heating appliances, and other consumer products that are significant sources of burn injuries, particularly products for which industry standards fail to give adequate protection.

40. The Commission recommends to schools giving degrees in architecture and engineering that they include in their curricula at least one course in fire safety. Further, we urge the American Institute of Architects, professional engineering societies, and State registration boards to implement this recommendation.

41. The Commission urges the Society of Fire Protection Engineers to draft model courses for architects and engineers in the field of fire protection engineering.

42. The Commission recommends that the proposed National Fire Academy develop short courses to educate practicing designers in the basics of fire safety design.

# CHAPTER 11

43. The Commission recommends that all local governmental units in the United States have in force an adequate building code and fire prevention code or adopt whichever they lack.

44. The Commission recommends that local governments provide the competent personnel, training programs for inspectors, and coordination among the various departments involved to enforce effectively the local building and fire prevention codes. Representatives from the fire department should participate in reviewing the fire safety aspects of plans for new building construction and alterations to old buildings.

45. The Commission recommends that, as the model code of the International Conference of Building Officials has already done, all model codes specify at least a single-station early-warning detector oriented to protect sleeping areas in every dwelling unit. Further, the model codes should specify automatic fire extinguishing systems and early-warning detectors for high-rise buildings and for low-rise buildings in which many people congregate.

## CHAPTER 12

46. The Commission recommends that the National Transportation Safety Board expand its efforts in issuance of reports on transportation accidents so that the information can be used to improve transportation fire safety.

47. The Commission recommends that the Department of Transportation work with interested parties to develop a marking system, to be adopted nationwide, for the purpose of identifying transportation hazards.

48. The Commission recommends that the proposed National Fire Academy disseminate to every fire jurisdiction appropriate educational materials on the problems of transporting hazardous materials.

49. The Commission recommends the extension of the Chem-Tree system to provide ready access by all fire departments and to include hazard control tactics.

50. The Commission recommends that the Department of the Treasury establish adequate fire regulations, suitably enforced, for the transportation, storage, and transfer of hazardous materials in international commerce.

51. The Commission recommends that the Department of Transportation set mandatory standards that will provide fire safety in private automobiles.

52. The Commission recommends that airport authorities review their firefighting capabilities and, where necessary, formulate appropriate capital improvement budgets to meet current recommended aircraft rescue and firefighting practices.

53. The Commission recommends that the Department of Transportation undertake a detailed review of the Coast Guard's responsibilities, authority, and standards relating to marine fire safety.

54. The Commission recommends that the railroads begin a concerted effort to reduce rail-caused fires along the Nation's rail system.

55. The Commission recommends that the Urban Mass Transportation Administration require explicit fire safety plans as a condition for all grants for rapid transit systems.

## CHAPTER 13

56. The Commission recommends that rural dwellers and others living at a distance from fire departments install early-warning detectors and alarms to protect sleeping areas.

57. The Commission recommends that U.S. Department of Agriculture assistance to [community fire protection facilities] projects be contingent upon an approved master plan for fire protection for local fire jurisdictions.

## CHAPTER 14

58. . . . the Commission recommends that the proposed U.S. Fire Administration join with the Forest Service, U.S.D.A., in exploring means to make fire safety education for forest and grassland protection more effective.

59. The Commission recommends that the Council of State Governments undertake to develop model State laws relating to fire protection in forests and grasslands.

60. The Commission urges interested citizens and conservation groups to examine fire laws and their enforcement in their respective States and to press for strict compliance.

61. The Commission recommends that the Forest Service, U.S.D.A., develop the methodology to make possible nationwide forecasting of fuel buildup as a guide to priorities in wildland management.

62. The Commission supports the development of a National Fire Weather Service in the National Oceanic and Atmospheric Administration and urges its acceleration.

## CHAPTER 15

63. The Commission recommends that the Department of Health, Education, and Welfare include in accreditation standards fire safety education in the schools throughout the school year. Only schools presenting an effective fire safety education program should be eligible for any Federal financial assistance.

64. The Commission recommends that the proposed U.S. Fire Administration sponsor fire safety education courses for educators to provide a teaching cadre for fire safety education.

65. The Commission recommends to the States the inclusion of fire safety education in programs educating future teachers and the requirement of knowledge of fire safety as a prerequisite for teaching certification.

66. The Commission recommends that the proposed U.S. Fire Administration develop a program, with adequate funding, to assist, augment, and evaluate existing public and private fire safety education efforts.

67. *. . . the Commission recommends that the proposed U.S. Fire Administration, in conjunction with the Advertising Council and the National Fire Protection Association, sponsor an all-media campaign of public service advertising designed to promote public awareness of fire safety.

68. The Commission recommends that the proposed U.S. Fire Administration develop packets of educational materials appropriate to each occupational category that has special needs or opportunities in promoting fire safety.

## CHAPTER 16

69. The Commission supports the Operation EDITH (Exit Drills In The Home) plan and recommends its acceptance and implementation both individually and community-wide.

70. The Commission recommends that annual home inspections be undertaken by every fire department in the Nation. Further, Federal financial assistance to fire jurisdictions should be contingent upon their implementation of effective home fire inspection programs.

71. The Commission urges Americans to protect themselves and their families by installing approved early-warning fire detectors and alarms in their homes.

72. The Commission recommends that the insurance industry develop incentives for policyholders to install approved early-warning fire detectors in their residences.

73. The Commission urges Congress to consider amending the Internal Revenue Code to permit reasonable deductions from income tax for the cost of installing approved detection and alarm systems in homes.

74. *. . . the Commission recommends that the proposed U.S. Fire Administration monitor the progress of research and development on early-warning detection systems in both industry and government and provide additional support for research and development where it is needed.

75. The Commission recommends that the proposed US. Fire Administration support the development of the necessary technology for improved automatic extinguishing systems that would find ready acceptance by Americans in all kinds of dwelling units.

76. The Commission recommends that the National Fire Protection Association and the American National Standards Institute jointly review the Standard for Mobile Homes and seek to strengthen it, particularly in such areas as interior finish materials and fire detection.

77. The Commission recommends that all political jurisdictions require compliance with the NFPA/ANSI standard for mobile homes together with additional requirements for early-warning fire detectors and improved fire resistance of materials.

78. The Commission recommends that State and local jurisdictions adopt the NFPA Standard on Mobile Home Parks as a minimum mode of protection for the residents of these parks.

# CHAPTER 17

79. The Commission strongly endorses the provisions of the Life Safety Code which require specific construction features, exit facilities, and fire detection systems in child day care centers and recommends that they be adopted and enforced immediately by all the States as a minimum requirement for licensing of such facilities.

80. The Commission recommends that early-warning detectors and total automatic sprinkler protection or other suitable automatic extinguishing systems be required in all facilities for the care and housing of the elderly.

81. The Commission recommends to Federal agencies and the States that they establish mechanisms for annual review and rapid upgrading of their fire safety requirements

for facilities for the aged and infirm, to a level no less stringent than the current NFPA Life Safety Code.

82. The Commission recommends that the special needs of the physically handicapped and elderly in institutions, special housing, and public buildings be incorporated into all fire safety standards and codes.

83. The Commission recommends that the States provide for periodic inspection of facilities for the aged and infirm, either by the State's fire marshal's office or by local fire departments, and also require approval of plans for new facilities and inspection by a designated authority during and after construction.

84. The Commission recommends that the National Bureau of Standards develop standards for the flammability of fabric materials commonly used in nursing homes with a view to providing the highest level of fire resistance compatible with the state-of-the-art and reasonable costs.

85. The Commission recommends that political subdivisions regulate the location of nursing homes and housing for the elderly and require that fire alarm systems be tied directly and automatically to the local fire department.

# CHAPTER 18

86. The Commission recommends that the Federal Government retain and strengthen its programs of fire research for which no non-governmental alternatives exist.

87. . . . the Commission recommends that the Federal budget for research connected with fire be increased by $26 million.

88. . . . the Commission recommends that associations of material and product manufacturers encourage their member companies to sponsor research directed toward improving the fire safety of the built environment.

# CHAPTER 19

89. . . . the Commission recommends that the proposed U.S. Fire Administration be located in the Department of Housing and Urban Development.

90. The Commission recommends that Federal assistance in support of State and local fire service programs be limited to those jurisdictions complying with the National Fire Data System reporting requirements.

# CHAPTER 20

No recommendations.

From November 30 through December 2, 1987, 14 years after *America Burning* was transmitted to the president, the U.S. Fire Administration conducted a three-day workshop to assess the U.S. fire problem and report on our nation's progress since the *America Burning* report. The participants included several who participated in the original work of the national Commission on Fire Prevention and Control, members of Congress, and officials from the U.S. Fire Administration and National Fire Academy. The findings of the work group, published as Appendix C of its report *America Burning Revisited*, are reproduced here. The entire text is available from the Government Printing Office or online from the U.S. Fire Administration in PDF format at www.usfa.fema.gov.

In 1999, James Lee Witt, director of the Federal Emergency Management Agency, recommissioned *America Burning* to assess our nation's progress in fire prevention and protection since the original report of 1973. The panel's report *America at Risk* is available from the Government Printing Office or online from the U.S. Fire Administration in PDF format at www.usfa.fema.gov.

## SECTION I   INTRODUCTION

One of the major events of the United States fire service occurred in June 1973—the publication of ***America Burning***. This document is the report of the National Commission on Fire Prevention and Control, a presidential commission appointed by President Richard M. Nixon. Richard E. Bland (associate professor, Pennsylvania State University) was commission chairman, and W. Howard McClennen (president of the International Association of Fire Fighters) was vice chairman. Commission members included members of Congress, the Secretaries of the Departments of Commerce and Housing and Urban development, and representatives of public and private organizations concerned with fire protection, including the fire service, insurance industry, news media, academia and the building industry.

The commission conducted research into the U.S. fire problem, held a series of hearings throughout the country and deliberated the solutions to these problems.

255

The result of this two-year effort was the commission's report, ***America Burning***.

This report contains 90 recommendations concerning the improvement of fire protection in the U.S. These recommendations cover the following general areas:

- burn prevention and treatment;
- fire fighter health and safety;
- building and fire codes and standards;
- automatic detection and suppression;
- fire protection master planning;
- fire department organization and operation;
- rural and wildland fire protection;
- public education;
- fire prevention inspection and enforcement;
- incentives for improved fire safety;
- transportation fire safety; and
- establishment of federal organizations for fire protection research, data collection and analysis, planning, and training.

In the 15 years since ***America Burning*** was published; many of the recommendations have been accomplished, and other recommendations have been accomplished partially. Some recommendations have been attempted without success, and others apparently have not been attempted at all.

The most visible accomplishment was the creation of the U.S. Fire administration, National Fire Academy and Center for Fire Research at the National Institute of Standards and Technology. However, some fire protection leaders have stated that this is only a partial success because the funding for these organizations has never come close to the amounts recommended by the commission.

Significant successes also have been achieved in such areas as installation of smoke detectors, improvement of fire and building codes, and fire protection master planning. Some success has been achieved for other recommendations, e.g., expanding public education programs and increasing fire department involvement in fire prevention. Other recommendations were not included in the legislation resulting from America Burning (P.L. 93-498) and, therefore, have not been implemented. For example, recommendations pertaining to fire department grants for training and equipment were not included in the legislation. Thus, such recommendations have not been accomplished at all.

The remainder of this report consists of the following sections:

Section II. Recommendation and Accomplishments—lists each of the 90 recommendations contained in America Burning along with a discussion of the extent to which each has been accomplished.

Section III. Summary of Accomplishments—contains a general discussion of (1) recommendations which have been substantially accomplished; (2) recommendations that have not been accomplished to any practical degree; and (3) recommendations for which significant progress has been made, but continued effort is still required.

## SECTION II  RECOMMENDATIONS AND ACCOMPLISHMENTS

## 2.1  Chapter 1 The Nation's Fire Problem

### Recommendation

1. The commission recommends that Congress establish a U.S. Fire Administration to provide a national focus for the nation's fire problem and to promote a comprehensive program with adequate funding to reduce life and property loss from fire.

### Accomplishments

The United States Fire Administration (USFA) was established in 1975 as the National Fire Prevention and Control Administration. Also established were the National Fire Academy (NFA), originally a unit of the fire administration, and the Center for Fire Research (CFR), a unit of the National Institute of Standards and Technology. The commission recommended an average annual budget for the first five years of $125 million. However, the largest annual budget ever received was less than $24 million, including funding for the CFR.

### Recommendation

2. The commission recommends that a national fire data system be established to provide a continuing review and analysis of the entire fire problem.

### Accomplishments

The National Fire Data Center (NFDC) was established in 1975 as a major component of the USFA. The National Fire Incident Reporting System (NFIRS) was implemented as the foundation of the NFDC, and now has 37 states and 20 metropolitan areas reporting data. Currently, the NFDC is conducting general data analysis, as well as special studies of fire fighter and residential fire facilities, and major and unusual fires. In addition, the NFDC is conducting a project to improve fire department long-range planning and tactical decision-making capabilities through automated management information systems.

## 2.2  Chapter 2 Living Victims of the Tragedy

### Recommendation

3. The commission recommends that Congress enact legislation to make possible the attainment of 25 burn units and centers and 90 burn programs within the next 10 years.

### Accomplishments

Not included in legislation (P.L. 93-498), thus, there are no corresponding programs. However, by 1983, there were approximately 125 burn centers in the United States, including one in virtually every metropolitan area with a sufficient population base, a significant achievement. Such federal actions as the enactment of Medicare and Medicaid, and support of medical research and training have contributed to this progress.

## Recommendation

4. The commission recommends that Congress, in providing for new burn treatment facilities, make adequate provision for the training and continuing support of the specialists to staff these facilities. Provision also should be made for the special training of those who provide emergency care for burn victims in general hospitals.

## Accomplishments

Not included in legislation (P.L. 93-498), thus, there are no corresponding programs. However, through other resources, this training is being carried out now on a broad scale by professional organizations and emergency medical service programs, and by burn centers throughout the nation.

## Recommendation

5. The commission recommends that the National Institutes of Health greatly augment their sponsorship of research on burns and burn treatment.

## Accomplishments

In the mid-1970s, the National Institute of General Medical Sciences granted funds for burn research programs at seven burn center hospitals and other public and private grants supported research at other burn centers.

## Recommendation

6. The commission recommends that the National Institutes of Health administer and support a systematic program of research concerning smoke inhalation injuries.

## Accomplishments

Not included in legislation (P.L. 93-498), thus, there are no corresponding programs. Also, see recommendation number 35.

# 2.3  Chapter 3 Are There Other Ways?

## Recommendation

7. The commission recommends that local governments make fire prevention at least equal to suppression in the planning of fire department priorities.

## Accomplishments

In general, fire prevention efforts have increased dramatically, including expanded bureau and in-service company inspections and enforcement, improved public education programs, and increased involvement in land use planning and building/fire code development. The USFA, NFPA and other fire protection organizations have conducted numerous projects to promote and support the expansion of fire prevention activities by local departments. While these projects have helped to increase the amount of local government resources devoted to fire prevention, the goal of making "fire prevention at least equal to suppression" is yet to be reached, except in a few departments. For example, the Visalia (California) Fire Department currently devotes approximately 60% of its annual budget to prevention.

## Recommendation

8. The commission recommends that communities train and use women for fire service duties.

## Accomplishments

The number of women in the fire service has increased significantly over the last 14 years. The first career woman fire fighter was hired by the Arlington County (Virginia) Fire Department in 1974 and is now a captain with that department. The USFA promoted the inclusion of women in the fire service and provided information to assist fire departments in recruiting and retaining women for fire fighting duties. This USFA program included the conduct of a conference on women in the fire service (1981), and preparation of the reports, Role of Women in the Fire Service, and Issues for Women in the Fire Service. The USFA is currently developing protective clothing and equipment sizing information for use by manufacturers in supplying items especially designed for female fire fighters.

## Recommendation

9. The commission recommends that laws which hamper cooperative arrangements among local fire jurisdictions be changed to remove the restrictions.

## Accomplishments

There are no known statistics on the number of mutual (including automatic) aid agreements that have been implemented over the past 14 years, or revision of associated laws. However, the use of such agreements has been promoted by the USFA (through its Integrated Emergency Management System [IEMS] project, for example), and by conducting a fire service seminar on mutual aid. It is felt that the number and extent of use of mutual aid agreements have increased. The budget limitations experienced by local governments probably have contributed to this increase.

## Recommendation

10. The commission recommends that every local fire jurisdiction prepare a master plan designed to meet the community's present and future needs in fire protection, to serve as a basis for program budgeting, and to identify and implement the optimum cost-benefit solutions in fire protection.

## Accomplishments

Numerous fire departments have prepared fire protection master plans as a result of the commission's recommendation and a comprehensive program conducted by the USFA. This program included the development of the master planning process (for both single and multi-jurisdictional planning), preparation of a set of manuals for use by departments in preparing plans, a national conference on master planning, a report to Congress on master planning, training courses and a planning support team. The number of departments that have prepared a master plan is unknown, but it is known that many plans have been prepared and some departments have updated their plans several times.

## Recommendation

11. The Commission recommends that federal grants for equipment and training be available only to those fire jurisdictions that operate from a federally approved master plan for fire protection.

## Accomplishments

The legislation (PL 93-498) did not provide for grants for training and equipment.

## Recommendation

12. The commission recommends that the proposed U.S. Fire Administration act as a coordinator of studies of fire protection methods and assist local jurisdictions in adapting findings to their fire protection planning.

## Accomplishments

The USFA has served as a clearinghouse for fire protection studies, ideas and information, and has disseminated such information through the Learning Resource Center and such publications as the arson and public education resource exchange bulletins.

## 2.4  Chapter 4 Planning for Fire Protection

## Recommendation

13. The commission recommends that the proposed U.S. Fire Administration provide grants to local fire jurisdictions for developing master plans for fire protection.

Further, the proposed U.S. Fire Administration should provide technical advice and qualified personnel to local fire jurisdictions to help them develop master plans.

## Accomplishments

The legislation (PL 93-498) did not provide for grants to fire departments for master planning. The USFA provided on-site master planning technical assistance during the period 1978–1982. This support was discontinued when the USFA was reorganized in 1983.

## 2.5  Chapter 5 Fire Service Personnel

### Recommendation

14. The commission recommends that the proposed U.S. Fire Administration sponsor research in the following areas: productivity measures of fire departments, job analyses, fire injuries and fire prevention efforts.

### Accomplishments

Numerous projects were conducted within these areas, especially fire fighter safety and fire prevention. The major productivity projects included a fire department working relationships study and the analysis of alternative company staffing levels. The apparatus staffing study was not completed because of USFA budget limitations. Currently, the USFA is conducting a study of alternative methods of providing fire protection, with results being prepared for dissemination to state and local governments.

### Recommendation

15. The commission urges the federal research agencies, for example, the National Science Foundation and the National Institute of Standards and Technology, to sponsor research appropriate to their respective missions within the areas of productivity of fire departments, causes of fire fighter injuries, effectiveness of fire prevention efforts and the skills required to perform various fire department functions.

### Accomplishments

The primary effort in these areas was accomplished by the USFA (see #14 above) because the missions of these agencies evolved into more technical issues. The USFA assumed responsibility for some of the fire projects initiated by NSF and NIST, for example, the pumping apparatus specifications and fire protection master planning projects. Projects conducted by the USFA in these areas resulted in the preparation of the following documents (as examples): Model Performance Criteria for Structural Fire Fighters' Helmets, Development of a Job-Related Physical Performance Examination for Fire Fighters— A Summary Report, Survey of Fire Fighter Injuries (in cooperation with IAFF), and

National Fire Service System. Task Analysis Phase I: Development of the Analysis Process. The USFA currently is sponsoring a fire fighter mortality study being conducted by the University of Washington.

## Recommendation

16. The commission recommends that the nation's fire departments recognize advanced and specialized education, and hire or promote persons with experience at levels commensurate with their skills.

## Accomplishments

Many fire departments have recognized that specialized education and advanced degrees can be of benefit to the department, and have included such criteria in recruiting or promoting for specific positions.

## Recommendation

17. The commission recommends a program of federal financial assistance to local fire services to upgrade their training.

## Accomplishments

The NFA currently provides financial assistance to local fire services through both resident and field training programs.

This assistance includes student stipends for resident classes, "train the trainer" (provides for training of state and local training personnel in the delivery of courses prepared for hand-off to state or local training organizations), direct supplementary delivery (supplements state and local training efforts), Academy Planning and Assistance Program (provides technical and training assistance to state and local fire service organizations), and Open Learning for the Fire Service (provides fire service personnel with the opportunity to earn baccalaureate degrees in fire administration or fire technology).

## Recommendation

18. In the administering of federal funds for training or other assistance to local fire departments, the commission recommends that eligibility be limited to those departments that have adopted an effective, affirmative action program related to the employment and promotion of members of minority groups.

## Accomplishments

This recommendation has been accomplished through the Federal Procurement Regulations because these regulations have equal opportunity requirements applicable to grantees.

## Recommendation

19. The commission recommends that fire departments, lacking emergency ambulance, paramedical and rescue services, consider providing them, especially if they are located in communities where these services are not provided adequately by other agencies.

## Accomplishments

Numerous fire departments have implemented and expanded EMS programs over the past 14 years. The USFA has conducted a number of activities to promote the establishment of fire service EMS programs, for example, the "National Workshop for Fire Service EMS Needs—the Rockville Report." In addition, the EMS standards established by DOT for personnel, vehicles and equipment have had a major impact on improving EMS programs at the local level. However, there are communities which still do not have adequate EMS programs.

# 2.6  Chapter 6 A National Fire Academy

## Recommendation

20. The commission recommends the establishment of a National Fire Academy to provide specialized training in areas important to the fire services and to assist state and local jurisdictions in their training programs.

## Accomplishments

The National Fire Academy has been operational since 1975 and has operated the fire training facility at Emmitsburg since January 1980. The academy has numerous programs to provide specialized training (hazardous materials, for example) and has provided assistance to local training activities, including course development and "train the trainer" courses.

## Recommendation

21. The commission recommends that the proposed National Fire Academy assume the role of developing, gathering and disseminating, to state and local arson investigators, information on arson incidents and on advanced methods of arson investigations.

## Accomplishments

The NFA conducts arson investigation and related training. The USFA maintains an arson research and information dissemination effort, with the fire data center collecting and analyzing information on arson incidents. Currently, the USFA is developing a community-based organization anti-arson program, a juvenile firesetter program and an Arson Information Management System, and is disseminating the results of a study of the needs of the rural arson investigator. The previously published Arson Resource Directory is being updated, and the Arson Resource Center, located at the Learning Resource Center, is now computerized for rapid access and updating. The CFR compiled the information, edited and published a Fire Investigation Handbook that is in widespread use among arson investigators.

## Recommendation

**22.** The commission recommends that the National Fire Academy be organized as a division of the proposed U.S. Fire Administration which would assume responsibility for deciding details of the academy's structure and administration.

## Accomplishments

During the period 1975–1981, the NFA was a component of the USFA. In 1981, the NFA was separated administratively from the USFA.

## Recommendation

**23.** The commission recommends that the full cost of operating the proposed National Fire Academy and subsidizing the attendance of fire service members be borne by the federal government.

## Accomplishments

The NFA pays up to 75% of the cost of attending courses at the academy. Generally, the only cost to the attendee is a nominal payment for meals and for local transportation at the departure location. This financial subsidy even includes airfare for attendees. However, continued subsidy is dependent on the availability of corresponding funds in the NFA budget.

## 2.7 Chapter 7 Equipping the Fire Fighter

### Recommendation

**24.** The commission urges the National Science Foundation, in its Experimental Research and Development Incentives Program, and the National Institute of Standards and Technology, in its Experimental Technology Incentives Program, to give high priority to the needs of the fire services.

### Accomplishments

The National Science Foundation stopped its applied fire research activities upon establishment of the USFA and CFR. The NIST Experimental Technology Incentives Program did conduct a major project on methods for fire retarding polyester/cotton apparel fabrics before the ETIP program was stopped. The CFR program has included several projects in response to the needs of the fire services, for example, fire fighter turnout coats, helmets, lightweight air tanks and methodology for locating fire stations.

## Recommendation

**25.** The commission recommends that the proposed U.S. Fire Administration review current practices in terminology, symbols and equipment descriptions, and seek to introduce standardization where it is lacking.

## Accomplishments

The USFA, NFA and CFR have promoted several areas of standardization, including the National Fire Incident Reporting System, fire fighter protective clothing and equipment, the Incident Command System, terminology and through training courses at the NFA.

## Recommendation

**26.** The commission urges rapid implementation of a program to improve breathing apparatus systems and expansion of the program's scope where appropriate.

## Accomplishments

Tremendous progress has been made in the design of safer, more effective breathing apparatus. In the early 1970s the NIST conducted a research and development project on the design of higher pressure, lighter weight air tanks. The USFA's Project FIRES, conducted in cooperation with the National Aeronautics and Space Administration (NASA), produced a new generation of breathing apparatus which is now widely available and used. The NFPA, under a USFA grant, prepared the Manual for Selection, Use, Care and Maintenance of Fire Fighter Self-Contained Breathing Apparatus. In addition, a USFA project with the Bureau of Mines addressed the design and testing of a (closed circuit) "Low Profile Rescue Breathing Apparatus." A prototype long-duration (two-hour), positive-pressure oxygen-breathing apparatus has been developed and is being field tested.

## Recommendation

**27.** The commission recommends that the proposed U.S. Fire Administration undertake a continuing study of fire service equipment needs, monitor research and development in progress, encourage needed research and development, disseminate results and provide grants to fire departments for equipment procurement to stimulate innovation in equipment design.

## Accomplishments

To date, the USFA's equipment research and development activities have been focused on fire fighter protective equipment, smoke detectors and reporting systems, automatic detection and suppression systems, and specifying pumping apparatus. All of these activities have resulted in the improvement of equipment design and operation, as well as the dissemination of corresponding information. The USFA and NASA have developed and are testing a "hands-free" communications device which is fully compatible with self-contained breathing apparatus. The USFA legislation did not provide for grants for equipment procurement. A new portable monitor that will quickly detect and identify hazardous chemical vapors is being field tested by fire departments under USFA sponsorship. The USFA is cooperating with the U.S. Coast Guard in the lab testing of chemical protective clothing, and with the Coast Guard and the Department of Energy in evaluating a hazardous chemical protective ensemble.

## Recommendation

28. The commission urges the Joint Council of National Fire Service Organizations to sponsor a study to identify shortcomings of fire fighting equipment and the kinds of research, development or technology transfer that can overcome the deficiencies.

## Accomplishments

Members of the joint council have contributed to numerous USFA and NFA activities, especially as members of project advisory committees and the NFA Board of Visitors. However, the specified study was not conducted. The USFA did conduct a project to establish fire protection research and development priorities.

## 2.8  Chapter 8 No Recommendations

## 2.9  Chapter 9 The Hazards Created Through Materials

## Recommendation

29. The commission recommends that research in the basic processes of ignition and combustion be strongly increased to provide a foundation for developing improved test methods.

## Accomplishments

The CFR conducts, both in-house and through a research grants program, fundamental and applied research in the physics and chemistry of fire leading to improved methods for measuring fire properties and fire performance of materials and products. The work ranges from flame chemistry and polymer degradation to the development of measurement and predictive methods. The fundamental research is used to underpin the more applied work. The research in flame chemistry is leading to the ability to predict the evolution of gases and particulates from the flame, e.g., soot and carbon monoxide. The work in polymer degradation is leading to the ability to design and produce more fire-safe materials. Measurement methods are being developed that will provide the scientifically based data needed as input to the predictive models.

## Recommendation

**30.** The commission recommends that the new Consumer Product Safety Commission (CPSC) give a high priority to the combustion hazards of materials in their end use.

## Accomplishments

The CPSC, with technical backup from the CFR, has addressed the fire hazard of many consumer products by establishing mandatory standards, promoting voluntary standards, urging product recalls and providing consumer information. Included products are children's sleep wear, general apparel, mattresses, upholstered furniture, blankets, fire-safe cigarettes, cigarette lighters, matches, fireworks and heating equipment.

## Recommendation

**31.** The commission recommends that the present fuel load study sponsored by the General Services Administration and conducted by the National Institute of Standards and Technology be expanded to update the technical study of occupancy fire loads.

## Accomplishments

The CPR conducted the fuel loads study for the GSA. The development by CFR of methods to predict the growth and spread of fire from measured fuel characteristics indicates the need for new surveys to provide information for fire hazards and risk calculations.

## Recommendation

**32.** The commission recommends that flammability standards for fabrics be given high priority by the Consumer Product Safety Commission.

## Accomplishments

The NIST and CPSC have implemented four flammable fabric standards, two which are mandatory and two which are voluntary in cooperation with manufacturers and trade associations. The children's sleepwear standard is mandatory. Prior to its adoption, there were approximately 60 child sleepwear-related fire deaths per year; now this number is approximately two deaths per year. A voluntary nightwear standard is being prepared in cooperation with industry. This standard will provide for point of sale comparative information on the flammability of various fibers. At the present time, approximately 75% of apparel fire deaths are persons over 66 years of age. This consumer information program should help to reduce that number. The Mattress Flammability Standard is mandatory and, as a result, almost every mattress being produced today will resist ignition by cigarette. The value of this standard is demonstrated by the fact that, during the period 1980–1984, there was an approximately 32% reduction in cigarette-ignited mattress fire deaths. In 1977, industry accepted a voluntary standard for upholstered furniture. Before this standard, approximately 10–15% of upholstered furniture would resist cigarette ignition; now approximately 68% will resist such ignition. During the period 1980–1984, cigarette-ignited upholstery fires decreased by approximately 24%.

## Recommendation

**33.** The commission recommends that all states adopt the Model State Fireworks Law of the National Fire Protection Association, thus prohibiting all fireworks except those for public displays.

## Accomplishments

The NFPA, USFA and other organizations have promoted the adoption of this law, and many states have adopted this or similar laws.

## Recommendation

**34.** The commission recommends that the Department of Commerce be funded to provide grants for studies of the dynamics of combustion and the means of its control.

## Accomplishments

The National Science Foundation RANN program on fire research was transferred to the CFR (a part of the National Institute of Standards and Technology which is an agency of the Department of Commerce) in the mid-1970s. CFR added funds from its base appropriation to increase the grants program to $2 million per year and annually funds more than 20 grants, mostly at universities. The CFR grants program is an integral part of the CFR program and is a way to bring the best scientific expertise in many disciplinary fields to focus on the fire problem. (See recommendation number 29 for a further description of the CFR program.)

## Recommendation

**35.** The commission recommends that the National Institute of Standards and Technology and the National Institutes of Health cooperatively devise and implement a set of research objectives designed to provide combustion standards for materials to protect human life.

## Accomplishments

CFR has conducted a broad program of research into such fire problem areas as ignition, flame spread, products of combustion, extinguishment, detection and heat release. This program has led to the establishment or modification of many standards, mostly through the consensus standards-setting process, for example, the CPSC children's sleepwear, mattress and insulation standards; the flooring radiant panel test for floor coverings adopted by ASTM, NFPA and ISO; testing and installation standards for smoke detectors adopted by NFPA, UL and others; and the flame spread test for wall lining materials adopted by the International Maritime Organization. However, standards have not specifically been developed in cooperation with the National Institutes of Health.

## 2.10  Chapter 10 Hazards Through Design

### Recommendation

36. The commission urges the National Institute of Standards and Technology to assess current progress in fire research and define the areas in need of additional investigation. Further, the institute should recommend a program for translating research results into a systematic body of engineering principles and, ultimately, into guidelines useful to code writers and building designers.

### Accomplishments

The CFR prepares both short- and long-term research plans based on an assessment of current and future needs of fire research results. In addition, CFR has established the National Fire Research Strategy Conference to prepare and continually update a fire research strategy for the nation. A major CFR thrust has been the translation of research results into a systematic body of engineering knowledge to provide guidelines, formulae, models, etc., to establish a soundly based fire protection engineering profession. Progress includes a set of engineering formulae known as FIRE-FORM, a prototype fire hazard assessment method called Hazard I, and several engineering models of varying sophistication that calculate fire growth and smoke spread.

### Recommendation

37. The commission recommends that the National Institute of Standards and Technology, in cooperation with the National Fire Protection Association and other appropriate organizations, support research to develop guidelines for a systems approach to fire safety in all types of buildings.

### Accomplishments

The systems approach to building fire safety continues as a major project in the CFR. An important series of developments were the Fire Safety Evaluation Systems for various occupancies adopted in consensus standards and/or in various regulations. A more scientifically based system is currently in development.

### Recommendation

38. The commission recommends that, in all construction involving federal money, awarding of those funds be contingent upon the approval of a fire safety systems analysis and a fire safety effectiveness statement.

## Accomplishments

A "Study of Fire Safety Effectiveness Statements" was conducted by the USFA as part of this effort, with the conclusion that liability issues would preclude the practical application of this concept. The General Services Administration now uses the fire safety systems analysis concept in designing new federal buildings.

## Recommendation

39. The commission urges the Consumer Product Safety Commission to give high priority to matches, cigarettes, heating appliances and other consumer products that are significant sources of burn injuries, particularly products for which industry standards fail to give adequate protection.

## Accomplishments

The CPSC, with technical support from the CFR, has addressed fire safety issues associated with matches, cigarettes, cigarette lighters and heating equipment, resulting in the adoption and use of mandatory requirements for matchbooks; the completion of a study of the technical and commercial feasibility of producing a cigarette that would not ignite material; the initiation of a cigarette lighter project which includes a study of how lighters start fires (approximately 200 people die each year from fires caused by lighters) and the consideration of lighter designs which are more difficult for use by children (i.e., "child proof"); and a study of electric, gas, kerosene and wood heaters to identify ways to improve design and maintenance. The CPSC has issued a regulation requiring labels on wood stoves which provides critical consumer information on installation, use and maintenance. (Approximately 20% of residential fires are caused by wood heating equipment.)

## Recommendation

40. The commission recommends to schools giving degrees in architecture and engineering that they include in their curricula at least one course in fire safety. Further, we urge the American Institute of Architects, professional engineering societies and state registration boards to implement this recommendation.

## Accomplishments

The CFR has been instrumental in establishing courses in fire science and fire safety engineering at several small colleges and universities. The referenced organizations have not implemented the recommendation, however.

## Recommendation

41. The commission urges the Society of Fire Protection Engineers to draft model courses for architects and engineers in the field of fire protection engineering.

## Accomplishments

The accomplishment in this area was achieved through a USFA grant to the Society of Fire Protection Engineers to prepare a methodology for fire-safe building design. The results of this project were documented in the report, Document the Final Fire Safety Methodology. Referenced model courses have yet to be developed.

## Recommendation

42. The commission recommends that the proposed National Fire Academy develop short courses to educate practicing designers in the basis of fire safety design.

## Accomplishments

The NFA has developed and offers resident, hand-off and Open Learning courses in fire-safe building design.

# 2.11  Chapter 11 Codes and Standards

## Recommendation

43. The commission recommends that all local governmental units in the United States have in force an adequate building code and fire prevention code or adopt whichever they lack.

## Accomplishments

The USFA and NFA have promoted the adoption of building/fire codes and the strengthening of existing codes. While quantitative data is not available, it is known that many communities have adopted new codes or improved existing codes. For example, many communities have adopted automatic detection and suppression system, smoke detector, fireworks and/or non-combustible roof ordinances. The NFA offers resident and field courses that address code development. There are still many communities that do not have adequate codes, and some do not have codes of any kind.

## Recommendation

44. The commission recommends that local governments provide the competent personnel, training programs for inspectors and coordination among the various departments involved to enforce effecively the local building and fire prevention codes. Representatives from the fire department should participate in reviewing the fire safety aspects of plans for new building construction and alterations to old buildings.

## Accomplishments

Building/fire code enforcement has improved significantly as fire departments have expanded their fire prevention programs (see Recommendation #7). Local and state governments (Florida, for example) have established training requirements for company and fire prevention bureau personnel who will be involved in inspection/enforcement activities. Relationships between fire and building departments have been enhanced in many communities, and, in some cases, the fire department has an inspector assigned to the building department. Furthermore, fire departments have become more involved in land use planning and review of proposed developments. The master planning program has helped in all of these areas because it stresses proactive fire protection activities. The USFA and NFA have promoted the improvement of inspection/enforcement capabilities through national conferences, manuals and training courses. The USFA recently has developed a new code implementation training program and has sponsored several related reports, including Management and Enforcement of Fire Codes, Administrative Aspects of Code Enforcement, and a four-volume set prepared by the American Bar Association, Alternatives for Effective Code Enforcement and Compliance Programs at the Local Level. The four volumes are for judges, prosecutors, local officials and code officials. To promote effective code enforcement, the USFA sponsored numerous meetings for city managers, fire chiefs and state governors. The NFA offers several courses which cover inspection and plan review, including "Fire Prevention Specialists I and II," and "Plans Review for Inspectors."

## Recommendation

45. The commission recommends that, as the model code of the International Conference of Building Officials has already done, all model codes specify at least a single-station, early-warning detector oriented to protect sleeping areas in every dwelling unit. Further, the model codes should specify automatic fire extinguishing systems and early-warning detectors for high-rise buildings in which many people congregate.

## Accomplishments

Significant advancement has been made in getting requirements for smoke detectors and automatic detection and suppression systems in codes. Currently, every major national model code requires smoke detectors in all new construction of dwelling units. In addition, some communities have amended these codes to require the retroactive installation of detectors in all dwelling units. The USFA and NFA have been extremely active in promoting the inclusion of requirements for detection and suppression systems in various codes. (The USFA provided significant input into the revision of the NFPA sprinkler standard [#13] and the development of the standard for residential sprinklers [#13D]). The USFA and CFR have conducted research into the effectiveness of such systems, held a number of conferences and workshops on these subjects, and prepared the National Directory of Automatic Suppression Systems and National Directory of Automatic Detection and Remote Alarm Systems, as well as numerous other reports which support the improvement of model codes in these areas. There are still many communities which do not have adequate codes for detection and suppression systems, especially for high-risk occupancies.

## 2.12  Chapter 12 Transportation Fire Hazards

### Recommendation

46. The commission recommends that the National Transportation Safety Board expand its efforts in issuance of reports on transportation accidents so that the information can be used to improve transportation fire safety.

### Accomplishments

The NTSB currently performs this function, for example, smoke detectors are required now in the cabins of all commercial passenger aircraft.

### Recommendation

47. The commission recommends that the Department of Transportation work with interested parties to develop a marking system, to be adopted nationwide, for the purpose of identifying transportation hazards.

### Accomplishments

The DOT placard system has been developed and implemented.

### Recommendation

48. The commission recommends that the proposed National Fire Academy disseminate to every fire jurisdiction appropriate educational materials on the problems of transporting hazardous materials.

### Accomplishments

The NFA offers on-campus courses in "Chemistry of Hazardous Materials" and "Hazardous Materials Tactical Considerations," and has developed hand-off hazardous materials training packages for use by state and local fire departments. However, there are still many fire service personnel (especially volunteers) who need, but have not received, such training. The NPA hazardous materials training program is expected to be expanded over the next three years as a result of the Superfund reauthorization legislation.

### Recommendation

49. The commission recommends the extension of the CHEMTREC system to provide ready access by all fire departments and to include hazard control tactics.

## Accomplishments

The CHEMTREC system has been expanded to provide increased information and assistance to fire departments. There is now a need to provide hazardous materials information using terminology that can be understood easily by fire service personnel (i.e., plain English).

## Recommendation

50. The commission recommends that the Department of the Treasury establish adequate fire regulations, suitably enforced, for the transportation, storage and transfer of hazardous materials in international commerce.

## Accomplishments

Regulations for the transportation, storage and transfer of hazardous materials, which would affect international commerce, have been implemented, but by the DOT and EPA rather than the treasury department.

## Recommendation

51. The commission recommends that the Department of Transportation set mandatory standards that will provide fire safety in private automobiles.

## Accomplishments

The DOT established a burn rate test requirement for interior materials in automobiles. Most of the fire safety standards for automobiles are voluntary.

## Recommendation

52. The commission recommends that airport authorities review their fire fighting capabilities and, where necessary, formulate appropriate capital improvement budgets to meet current recommended aircraft rescue and fire fighting practices.

## Accomplishments

The intent of this recommendation has been accomplished by the DOT.

## Recommendation

53. The commission recommends that the Department of Transportation undertake a detailed review of the Coast Guard's responsibilities, authority and standards relating to marine fire safety.

## Accomplishments

This is an on-going Coast Guard effort.

## Recommendation

**54.** The commission recommends that the railroads begin a concerted effort to reduce rail-caused fires along the nation's rail system.

## Accomplishments

Progress has been made in this area through the use of hot-box detectors and spark arresters.

## Recommendation

**55.** The commission recommends that the Urban Mass Transportation Administration require explicit fire safety plans as a condition for all grants for rapid transit systems.

## Accomplishments

Grants for rapid transit systems now include fire safety plan requirements.

# 2.13  Chapter 13 Rural Fire Protection

## Recommendation

**56.** The commission recommends that rural dwellers and others living at a distance from fire departments install early-warning detectors and alarms to protect sleeping areas.

## Accomplishments

The USFA and other fire organizations have conducted extensive public education programs to encourage the use of detection and warning systems in rural areas.

## Recommendation

**57.** The commission recommends that U.S. Department of Agriculture assistance to community fire protection facilities projects be contingent upon an approved master plan for fire protection for local jurisdictions.

## Accomplishments

This recommendation has not been implemented by the Department of Agriculture. However, the USFA has developed and promoted a master planning process designed especially for small towns and rural areas, the Basic Guide for Fire Prevention and Control Master Planning.

# 2.14  Chapter 14 Forest and Grassland Fire Protection

## Recommendation

**58.** The commission recommends that the proposed U.S. Fire Administration join with the U.S. Forest Service in exploring means to make fire safety education for forest and grassland protection more effective.

## Accomplishments

The USFA has been working with the U.S. Forest Service since 1975, and an active program is underway. Furthermore, the USFA is a member of the National Wildfire Coordinating Group.

## Recommendation

**59.** The commission recommends that the Council of State Governments should develop model state laws relating to fire protection in forests and grasslands.

## Accomplishments

This recommendation has not been initiated.

## Recommendation

**60.** The commission urges interested citizens and conservation groups to examine fire laws in their respective states and to press for strict compliance.

## Accomplishments

Individuals and organizations have been performing this function (lobbying for fire-safe roofing laws, for example). However, these efforts have not been coordinated at the national level.

## Recommendation

**61.** The commission recommends that the U.S. Forest Service develop the methodology to make possible nationwide forecasting of fuel build-up as a guide to priorities in wildland management.

## Accomplishments

The USFS has an active fuel management program in progress which includes pre-deployment of resources as a function of fuel loading, weather and other factors.

## Recommendation

**62.** The commission supports the development of a National Fire Weather Service in the National Oceanic and Atmospheric Administration and urges its acceleration.

## Accomplishments

This recommendation has been accomplished; the National Fire Weather Service is operational.

## 2.15  Chapter 15 Fire Safety Education

### Recommendation

**63.** The commission recommends that the Department of Health, Education and Welfare (now the Department of Health and Human Services) include in accreditation standards fire safety education in the schools throughout the school year. Only schools presenting an effective fire safety education program should be eligible for any federal financial assistance.

### Accomplishments

Some states and local school districts have adopted requirements for fire safety education. However, it is not known if the Department of Education has established accreditation standards.

### Recommendation

**64.** The commission recommends that the proposed U.S. Fire Administration sponsor fire safety education courses for educators to provide a teaching cadre for fire safety education.

### Accomplishments

The USFA has actively promoted fire safety education through conferences and preparation of manuals and materials. For example, the USFA prepared the Public Fire Education Planning Manual, and the report, Young Children as New Targets for Public Fire Education. The NFA conducts training courses to help in working with educators and the school system, for example, "Introduction to Fire Safety Education" and "Advanced Fire Safety Education."

### Recommendation

**65.** The commission recommends to the states the inclusion of fire safety education in programs for future teachers and the requirement of fire safety knowledge as a prerequisite for teaching certification.

### Accomplishments

The degree to which states have implemented this recommendation is unknown, which probably means that little has been accomplished.

### Recommendation

**66.** The commission recommends that the proposed U.S. Fire Administration develop 'a program, with adequate funding, to assist, augment, and evaluate existing public and private fire safety education efforts.'

## Accomplishments

Since 1975, the USFA has conducted a program to support local public education activities. The extent of this program has varied according to the extent of funding available each fiscal year. This program has included the general promotion of public education efforts by local agencies through meetings (e.g., the "Public Fire Education Planning Conference"); preparation of support materials (e.g., Media Ideas Workbook); and provision of limited financial support through the "Public Education Assistance Program" (PEAP) which was offered during the period 1979-1981 and had to be discontinued as a result of funding reductions. The USFA is sponsoring the Community Volunteer Fire Prevention Program to promote the cooperation of local fire departments and citizen groups in conducting community fire prevention, education and protection programs.

## Recommendation

67. The commission recommends that the proposed U.S. Fire Administration, in conjunction with the National Advertising Council and National Fire Protection Association, sponsor an all-media campaign of public service advertising designed to promote public awareness of fire safety.

## Accomplishments

The USFA has conducted a number of national campaigns (e.g., the national smoke detector installation and maintenance campaign). In fiscal year 1985, a renewed public education and awareness effort was initiated as a result of additional Congressional support. An all-media campaign conducted in conjunction with the National Advertising Council has not been accomplished. However, such a program still is being pursued.

## Recommendation

68. The commission recommends that the proposed U.S. Fire Administration develop packets of educational materials appropriate to each occupational category that has special needs or opportunities in promoting fire safety.

## Accomplishments

Although packets have not been developed for specific occupations, materials for preschool children and teachers were developed in cooperation with the Children's Television Workshop (i.e., "Sesame Street"). The award winning Sesame Street Fire Safety Program is being expanded to include older children.

## 2.16  Chapter 16 Fire Safety for the Home

### Recommendation

**69.** The commission supports the Operation EDITH (Exit Drills In The Home) plan and recommends its acceptance and implementation both individually and communitywide.

### Accomplishments

The NFPA has made a significant effort in accomplishing this recommendation. The EDITH program is accepted and used widely by local fire service and community groups.

### Recommendation

**70.** The commission recommends that annual home inspections be undertaken by every fire department in the nation. Further, federal financial assistance to fire jurisdictions should be contingent upon their implementation of effective home fire inspection programs.

### Accomplishments

Home fire inspection programs of various types are conducted by most fire departments. The USFA has encouraged the initiation of such programs through research (e.g., Project RIDFIRE and the Municipal Fire Insurance Analysis): conferences (e.g., "Dynamics of Fire Prevention"); and preparation of materials (e.g., the Edmonds [Washington] Home Survey materials). Because of the constitutional issue of eminent domain, a more direct federal role is not considered appropriate.

### Recommendation

**71.** The commission urges Americans to protect themselves and their families by installing approved early-warning fire detectors and alarms in their homes.

### Accomplishments

As a result of vigorous programs involving the USFA, other federal agencies, national fire service organizations, and state and local fire protection agencies, smoke detectors are now in an estimated 80% of dwelling units nationwide. There is now a need for programs to ensure that these detectors are tested and maintained properly.

### Recommendation

**72.** The commission recommends that the insurance industry develop incentives for policyholders to install approved early-warning fire detectors in their residences.

## Accomplishments

Many residential fire insurance policies now offer discounts for use of smoke detectors and residential sprinkler systems.

## Recommendation

**73.** The commission urges Congress to consider amending the Internal Revenue Code to permit reasonable deductions from income tax for the cost of installing approved detection and alarm systems in homes.

## Accomplishments

The USFA actively pursued this recommendation. However, the IRS did not feel that the income tax could be used as a fire safety incentive. Furthermore, a sponsor was not identified for the necessary legislation. However, in a related area, California adopted a constitutional amendment excluding the value of newly installed automatic detection and suppression systems in establishing property value for taxing purposes.

## Recommendation

**74.** The commission recommends that the proposed U.S. Fire Administration monitor the progress of research and development on early-warning detection systems in both industry and government and provide additional support for research and development where needed.

## Accomplishments

The USFA and CFR have been involved extensively in smoke detector research and development. Examples of corresponding publications include New Concepts of Fire Detection, Analysis of Fire Detector Test Methods/Performance, and Detector Sensitivity and Siting Requirements for Dwellings. The CFR is studying the detector false alarm problem in hospitals.

## Recommendation

**75.** The commission recommends that the proposed U.S. Fire Administration support the development of the necessary technology for improved automatic extinguishing systems that would find ready acceptance by Americans in all kinds of dwelling units.

## Accomplishments

As a result of USFA activities, significant progress has been made in the development and installation of residential sprinkler systems. These activities have included support of system design (e.g., Development of Low-Cost Residential Sprinkler Protection: A

Technical Report, Study to Establish the Existing Automatic Fire Suppression Technology for Use in Residential Occupancies, and Sprinkler Performance in Residential Fire Tests). In addition, the USFA was a sponsor of "Operation San Francisco" which was a series of tests of retrofitted sprinkler systems (including a residential occupancy) using advanced designs and materials. The USFA also has been active in sponsoring a number of meetings to disseminate the results of these efforts and encourage the implementation of such systems. For example, the USFA currently is sponsoring a nationwide series of sprinkler workshops in cooperation with the International Association of Fire Chiefs.

## Recommendation

76. The commission recommends that the National Fire Protection Association and American National Standards Institute jointly review the Standard for Mobile Homes and seek to strengthen it, particularly in such areas as interior finish materials and fire detection.

## Accomplishments

The CFR, after considerable research, recommended guidelines for fire safety to HUD, directed primarily at interior finish, for the agency's minimum property standards for mobile homes. These guidelines were adopted and implemented.

## Recommendation

77. The commission recommends that all political jurisdictions require compliance with the NFPA/ANSI standard for mobile homes, together with additional requirements for early-warning fire detectors and improved fire resistance of materials.

## Accomplishments

As a result of NFPA activities, mobile home standards now require the installation of smoke detectors in new mobile homes, as well as exiting requirements and interior flame spread restrictions.

## Recommendation

78. The commission recommends that state and local jurisdictions adopt the NFPA Standard on Mobile Home Parks as a minimum mode of protection for the residents of these parks.

## Accomplishments

Many jurisdictions have adopted the NFPA standard.

## 2.17  Chapter 17 Fire Safety for the Young, Old and Infirm

### Recommendation

79. The commission strongly endorses the provisions of the Life Safety Code which require specific construction features, exit facilities and fire detection systems in child day care centers, and recommends that they be adopted and enforced immediately by all the states as a minimum requirement for the licensing of such facilities.

### Accomplishments

Many states and communities have adopted NFPA's Life Safety Code.

### Recommendation

80. The commission recommends that early-warning detectors and total automatic sprinkler protection or other suitable automatic extinguishing systems be required in all facilities for the care and housing of the elderly.

### Accomplishments

All national codes and state statutes now require such protection for congregate care facilities.

### Recommendation

81. The commission recommends to federal agencies and the states that they establish mechanisms for annual review and rapid upgrading of their fire safety requirements for facilities for the aged and infirm to a level no less stringent than the current NFPA Life Safety Code.

### Accomplishments

The code development process provides for the accomplishment of this recommendation.

### Recommendation

82. The commission recommends that the special needs of the physically handicapped and elderly in institutions, special housing, and public buildings be incorporated into all fire safety standards and codes.

### Accomplishments

This recommendation must be accomplished by standard-setting and code organizations. The USFA has analyzed these problems (e.g., Fire and Life Safety for the Handicapped) and has encouraged improvement of standards and codes and adoption by state and local governments. The NFPA has addressed the special needs of congregate care facilities in the Life Safety Code (101). If federal funds are involved, such facilities must meet the codes and standards.

## Recommendation

83. The commission recommends that the states provide for periodic inspection of facilities for the aged and infirm, either by the state fire marshal's office or by local fire departments, and also require approval of plans for new facilities and inspection by a designated authority during and after construction.

## Accomplishments

Many states and local jurisdictions now have such requirements. The USFA has encouraged the adoption of these requirements.

## Recommendation

84. The commission recommends that the National Institute of Standards and Technology develop standards for the flammability of fabric materials commonly used in nursing homes, with a view to providing the highest level of fire resistance compatible with the state-of-the-art and reasonable costs.

## Accomplishments

The CPSC, with technical support from CFR, has established several standards under the Flammable Fabrics Act that are also applicable to the needs of nursing homes. In addition, CFR developed the Fire Safety Evaluation System for health care facilities that aids in the objective fire safety evaluation of such facilities by both the fire service and owner/operators.

## Recommendation

85. The commission recommends that political subdivisions regulate the location of nursing homes and housing for the elderly and require that fire alarm systems be tied directly and automatically to the local fire department.

## Accomplishments

This recommendation would be implemented through local zoning regulations and codes. Many communities now require central station monitoring of automatic detection and suppression systems in these occupancies.

## 2.18  Chapter 18 Research for Tomorrow's Fire Problem

### Recommendation

86. The commission recommends that the federal government retain and strengthen its programs of fire research for which no non-governmental alternatives exist.

### Accomplishments

Fire research programs have been conducted by the USFA and CFR, but not to the extent envisioned in America Burning because of budget limitations. For example, the commission recommended an annual research budget (USFA and CFR) of $33,250,000 (1973 dollars). The largest budget ever given to the USFA for all programs was approximately $24 million.

### Recommendation

87. The commission recommends that the federal budget for research connected with fire be increased by $26 million.

### Accomplishments

See recommendation 86.

### Recommendation

88. The commission recommends that associations of material and product manufacturers encourage their member companies to sponsor research directed toward improving the fire safety of the built environment.

### Accomplishments

Associations and manufacturers have increased fire research activities, for example, the Chemical Manufacturers Association, The Society of the Plastics Industry, Carpet and Rug Institute, Tobacco Institute, and Concrete and Masonry Institute. These associations also have sponsored research at such private laboratories as the Southwest Research Institute, Factory Mutual and Underwriters Laboratories.

## 2.19  Chapter 19 Federal Involvement

### Recommendation

89. The commission recommends that the proposed U.S. Fire Administration be located in the Department of Housing and Urban Development.

## Accomplishments

Originally, the USFA was in the Department of Commerce and was moved to the Federal Emergency Management Agency in 1979.

## Recommendation

**90.** The commission recommends that federal assistance in support of state and local fire service programs be limited to those jurisdictions complying with the National Fire Data System reporting requirements.

## Accomplishments

The federal assistance funding categories included in the commission's proposed budget were not included in subsequent U.S. Forest Service funding. Therefore, this recommendation is not applicable.

# SECTION III   SUMMARY OF ACCOMPLISHMENTS

The recommendations which have been fully or substantially accomplished generally have been associated with the agencies created as a result of America Burning or those with a public safety mission such as the Consumer Product Safety Commission.

Conversely, those recommendations not accomplished were generally the responsibility of agencies not primarily concerned with fire safety, for example, the U.S. Department of Agriculture, Department of Education and the Council of State Governments.

## 3.1  Recommendations Generally Accomplished

The recommendations that are fully or practically accomplished are listed below:

1. The commission recommends that Congress establish a U.S. Fire Administration to provide a national focus for the nation's fire problem and to promote a comprehensive program with adequate funding to reduce life and property loss from fire. (Note; The USFA was established, but never funded at the recommended levels.)
18. In the administering of federal funds for training or other assistance to local fire departments, the commission recommends that eligibility be limited to those departments that have adopted an effective affirmative action program related to the employment and promotion of members of minority groups.
20. The commission recommends the establishment of a National Fire Academy to provide specialized training in areas important to the fire services and to assist state and local jurisdictions in their training programs. (Note: The NFA was established, but never funded at the recommended levels.)
42. The commission recommends that the proposed National Fire Academy develop short courses to educate practicing designers in the basis of fire safety design.

45. The commission recommends that, as the model code of the International Conference of Building Officials has already done, all model codes specify at least a single-station early-warning detector oriented to protect sleeping areas in every dwelling unit. Further, the model codes should specify automatic fire extinguishing systems and early-warning detectors for high-rise buildings in which many people congregate.

46. The commission recommends that the National Transportation Safety Board expand its efforts in issuance of reports on transportation accidents so that the information can be used to improve transportation fire safety.

47. The commission recommends that the Department of Transportation work with interested parties to develop a marking system, to be adopted nationwide, for the purpose of identifying transportation hazards.

49. The commission recommends the extension of the CHEMTREC system to provide ready access by all fire departments and to include hazard control tactics.

50. The commission recommends that the Department of the Treasury establish adequate fire regulations, suitably enforced, for the transportation, storage and transfer of hazardous materials in international commerce.

52. The commission recommends that airport authorities review their fire fighting capabilities and, where necessary, formulate appropriate capital improvement budgets to meet current recommended aircraft rescue and fire fighting practices.

55. The commission recommends that the Urban Mass Transportation Administration require explicit fire safety plans as a condition for all grants for rapid transit systems.

62. The commission supports the development of a National Fire Weather Service in the National Oceanic and Atmospheric Administration and urges its acceleration.

77. The commission recommends that all political jurisdictions require compliance with the NFPA/ANSI standard for mobile homes, together with additional requirements for early-warning fire detectors and improved fire resistance of materials.

80. The commission recommends that early-warning detectors and total automatic sprinkler protection or other suitable automatic extinguishing systems be required in all facilities for the care and housing of the elderly.

81. The commission recommends to federal agencies and the states that they establish mechanisms for annual review and rapid upgrading of their fire safety requirements for facilities for the aged and infirm to a level no less stringent than the current NFPA Life Safety Code.

## 3.2  Recommendations Not Accomplished

The recommendations without any significant accomplishments are listed below:

6. The commission recommends that the National Institutes of Health administer and support a systematic program of research concerning smoke inhalation injuries.

11. The commission recommends that federal grants for equipment and training be available only to those fire jurisdictions that operate from a federally approved master plan for fire protection.

28. The commission urges the Joint Council of National Fire Service Organizations to sponsor a study to identify shortcomings of fire fighting equipment and the kinds of research, development or technology transfer that can overcome the deficiencies.

40. The commission recommends to schools giving degrees in architecture and engineering that they include in their curricula at least one course in fire safety. Further, we urge the American Institute of Architects, professional engineering societies and state registration boards to implement this recommendation.

41. The commission urges the Society of Fire Protection Engineers to draft model courses for architects and engineers in the field of fire protection engineering.

57. The commission recommends that U.S. Department of Agriculture assistance to community fire protection facilities projects be contingent upon an approved master plan for fire protection for local jurisdictions.

59. The commission recommends that the Council of State Governments undertake to develop model state laws relating to fire protection in forests and grasslands.

73. The commission urges Congress to consider amending the Internal Revenue Code to permit reasonable deductions from income tax for the cost of installing approved detection and alarm systems in homes.

87. The commission recommends that the federal budget for research connected with fire be increased by $26 million.

89. The commission recommends that the proposed US. Fire Administration be located in the Department of Housing and Urban Development.

90. The commission recommends that federal assistance in support of state and local fire service programs be limited to those jurisdictions complying with the National Fire Data System reporting requirements.

## 3.3  Recommendations Partially Accomplished

All of the remaining recommendations fall into this category. Each recommendation has been accomplished to some degree, but additional effort is required. In some cases, the recommendation may never be accomplished fully because it is on-going and general in nature.

In many cases, recommendations could not be completed because of a lack of resources, especially considering that the USFA, NFA and CFR were never funded at anywhere near the levels recommended by the commission.

## 3.4  Conclusions

The accomplishments discussed in this report are truly significant. In fact, 79 of the 90 recommendations have been accomplished to some degree, and the consequences of these accomplishments have had a major impact on the protection of life and property. For example:

- The annual number of fire deaths has decreased by 23% from 1975 to 1985 (an average annual reduction of 1900 deaths). On a cumulative basis, this reduction means that an estimated 6,900 lives have been saved between 1975 and 1985. Firefighter deaths also have been reduced significantly.

- Even greater reductions in fire deaths have been achieved within special categories. Clothing fire deaths have fallen by 73% over the period 1968–1983. In addition, children's clothing fire deaths have dropped by 90%.

- The number of fires reported to fire departments has decreased by approximately 20%) over the period 1975–1985. The number of reported fires has been decreasing even though the population is increasing which means that the number of fires is declining even on a per capita basis.

These achievements demonstrate what has been accomplished since America Burning was published, and the potential for what could be accomplished if the remaining recommendations were substantially completed.

# GLOSSARY

## A

**Acceptable risk** The level of fire risk that the general public is willing to bear at a given time.

**Accreditation** The act of assessing the fitness of an organization to test and certify individuals in accordance with a standard. Accredited agencies are authorized to test and certify individuals.

**Active fire protection feature** A fire protection feature or system such as automatic sprinklers, fire detectors, or smoke removal systems that operate or activate automatically or manually.

**Adopted budget** Finalized budget that has been formally approved by the governing body of the jurisdiction.

**Adoption** The acceptance and formal approval of a model code by the governing body of a jurisdiction.

**Adoption by reference** Code adoption process in which the jurisdiction passes an ordinance that lists or references a specific edition of a model code.

**Adoption by transcription** Code adoption process in which the model code is republished as an ordinance by a jurisdiction.

## B

**Budget** Government financial plan that includes spending and income, often divided by categories for a given time, normally one year.

**Building Code Effectiveness** A system developed by ISO to evaluate the potential effectiveness of local government building regulatory systems, patterned after the public protection classification system (fire department grading schedule).

## C

**Capital improvement budget** Long-range government financial plan typically used to fund construction projects, road construction, and land acquisition.

**Certification** Approval by an accredited agency that an individual meets the requirements of a standard, accompanied by the issuance of a certificate.

**Chain of custody** System to verify by documentation that evidence has been in the sole control of law enforcement from the time it is seized to the time it is admitted in court.

**Civilian employees** Fire department employees in a support position who are not trained firefighters.

**Code** A systematically arranged body of rules. When and where to do or not to do something.

**Conflagration** A very large destructive fire that defies control and causes extensive damage over a large area.

**Construction classification** Classification of a building or structure into one of the five

289

types or subtypes included in the model building codes based on types of materials and structural protection afforded.

## D

**Defend-in-place strategy**  Design system used for structures housing persons in which the likelihood of evacuation is remote because of infirmity, disability, confinement, or age; it includes fire protection features such as fire resistance-rated construction and automatic sprinklers, intended to extend tenable conditions during extinguishment or rescue.

## E

**Expert witness**  A person who by education or specialized experience possesses superior knowledge and is permitted to offer opinions during court testimony.

## F

**Fire brigade**  A private firefighting force within an industrial or government complex, usually employees of the facility, trained and equipped for on site firefighting.

**Fire Command Centers**  Fire resistance-rated rooms normally found within high-rise buildings and other large structures that contain controls for fire protection systems, building systems and utilities, and communications systems.

**Fire department connections**  External hose connections to supply fire protection water for automatic sprinklers and standpipes.

**Fire protection ratings**  Protection provided for building elements from the effects of fire expressed in terms of time.

**Fire protection system**  A system that detects fire or combustion products, suppresses or extinguishes fire, retards the passage of fire or smoke, or makes notification or alarm.

**Fire-resistance–rated assemblies**  Assemblies of materials designed and tested to retard the passage of heat, smoke, and fire for a given period of time.

**Fiscal year**  Budget cycle year, usually July 1 through June 30. The federal fiscal year is October 1 through September 30.

**Freedom of Information and Public Access laws**  Federal or state laws that provide for public access to government documents and meetings.

## I

**Incendiary**  A destructive fire that is intentionally set.

**Inspection model**  Method of determining inspection priorities in which local officials select the occupancies to be inspected.

**Inspection warrant**  Administrative warrant. A warrant issued by a court with jurisdiction commanding an officer to inspect a specific premises.

**Institutional occupancy**  An occupancy that houses persons whose capacity for self-preservation is limited or diminished by reason of age, infirmity, or confinement.

**Interior finish**  The exposed interior surface of a building or structure, whether for acoustical, decorative, insulative, or fire protection purposes.

## J

**Job performance requirements (JPRs)**  A statement within a certification standard that describes a specific job task, lists the items necessary to complete the task, and defines measurable or observable outcomes and evaluation areas.

## K

**Knowledge, skills, and abilities (KSAs)**  Minimum knowledge and attendant skills and ability needed to perform a job.

## L

**Line item**  Fund category within a budget.

## M

**Mini-maxi code**  Code adopted at a state level that cannot be locally amended.

**Model codes**  A code developed by an organization for adoption by governments.

**Mutual fire insurance**  Not-for-profit system in which all policyholders are members of the company. When premiums exceed losses, surplus funds are distributed among the members.

**N**

**Notice of violation**  Written notice issued to a property owner or occupant listing (1) unsafe conditions, (2) applicable code sections, (3) required corrective action, and (4) date of a follow-up inspection to ensure compliance.

**O**

**Operating budget**  Government financial plan that includes funding for the day-to-day operations of government, including salaries, utilities, rent, fuel, and transfer payments.

**Optical character recognition (OCR)**  Software that "reads" written documents when scanned and digitizes characters.

**Ordinance**  Law of a political subdivision of a state.

**Outside screw and yoke valve**  A fire protection system water supply valve in which the stem protrudes from the housing when the valve is shut, making the valve condition readily apparent.

**P**

**Passive fire protection**  Built-in fire protection features such as rated construction that provide fire safety and do not require activation.

**Permit**  Approval from the appropriate code official that authorizes construction, operation of a regulated process, or the maintenance of a regulated occupancy type.

**Permit conditions**  Code-specified conditions regarding permit issuance, including posting of the permit, submitting to inspection, accuracy in the permit application process, and complying with all code provisions.

**Permit system**  System of determining inspection priorities based on permits required by the model fire codes.

**Preblast survey**  Inspection of the structures in the vicinity of future blasting operations. The surveys identify existing structural defects and damage and give advance notice to surrounding property owners.

**Professional registration**  State-required licensure system for professionals such as engineers and architects.

**Public Protection Classification System**  The ISO's system of rating a jurisdiction's system of public fire protection, including the fire department, water supply, and communications system.

**Public record**  A record, memorial of some act or transaction, written evidence of something done, or document considered as either concerning or interesting the public or open to public inspection; includes all documents: papers, letters, maps, books, photographs, films, sound recordings, magnetic or other tapes, electronic data-processing records, artifacts, or other documentary material, regardless of physical form or characteristics.

**R**

**Recommended budget**  Proposed budget submitted to the legislative body by the executive for action. When approved after public hearings, it becomes the approved budget.

**Right of entry clause**  Clause within the model fire codes that authorizes the fire official to enter buildings and premises for the purpose of enforcing the code, with the permission of the occupant.

**S**

**Seismograph records**  Records of ground vibration and airblast measured during blasting operations.

**Selective code enforcement**  Illegal code enforcement actions based on political or other motives.

**Shall**  Indicates a positive and definitive requirement of the code that must be performed; action is mandatory.

**Shot records**  Records of blasting operations.

**Site plan**  A plan for proposed development. Typically the first plan submitted to the jurisdiction that includes the location of structures, occupancy and construction type, proposed roadway and parking facilities, and available fire flow.

**Special amusement buildings**  A permanent or temporary structure that provides a walkway or method of transportation, wherein the required means of egress is not apparent or is confounded through the use of theatrical or special effects.

**Standard** A rule for measuring or a model to be followed. How to do something, what materials to use. Also known as referenced standard.

**Standpipe** A system of piping, valves hose outlets, and allied equipment installed in a building to distribute fire protection water.

**Stock fire insurance** Fire insurance provided by commercial, for-profit companies.

**Structural fire protection** Protection afforded to structural elements to resist the effects of fire. Protection is generally provided through encasement with concrete, gypsum, or other approved materials.

## T

**Technical codes** Codes designed to regulate technical processes such as construction; installation of electrical, mechanical, and plumbing systems; hazardous industrial processes; and building electrical, mechanical, plumbing, and property maintenance codes.

**Third-party agency** Nongovernmental agency that has been approved by the code official to conduct tests and inspections. Test and inspection results are forwarded to a code official who approves or rejects the installation, construction, or operation.

**Third-party inspection** Inspection or testing by an approved nongovernmental agency that is forwarded to the code official for action or an inspection performed on behalf of the local government.

**Trade-off incentive** Code provisions that permit the reduction of fire protection ratings and increases in building height, area, number of stories, and reduction in means of egress provisions in return for the installation of sprinkler systems.

**Trade secrets** Propriety information regarding a business or manufacturing process.

## U

**Uniformed employees** Trained firefighters within a fire department.

**User fee** Fee for service charged by a government organization for providing a specific service.

# INDEX

*Note:* Page numbers followed by an *f* indicate figures; numbers followed by a *n* indicate a Note.